Transport Systems and Processes

Marine Navigation and Safety of Sea Transportation

Editors

Adam Weintrit & Tomasz Neumann
Gdynia Maritime University, Gdynia, Poland

CRC Press
Taylor & Francis Group
Boca Raton London New York Leiden

CRC Press is an imprint of the
Taylor & Francis Group, an **informa** business

A BALKEMA BOOK

CRC Press/Balkema is an imprint of the Taylor & Francis Group, an informa business

© 2011 Taylor & Francis Group, London, UK

Printed and bound in Great Britain by Antony Rowe Ltd (A CPI-group Company),
Chippenham, Wiltshire

Published by: CRC Press/Balkema
P.O. Box 447, 2300 AK Leiden, The Netherlands
e-mail: Pub.NL@taylorandfrancis.com
www.crcpress.com – www.taylorandfrancis.co.uk – www.balkema.nl

ISBN: 978-0-415-69120-8 (Pbk)
ISBN: 978-0-203-15700-8 (eBook)

List of reviewers

Prof. Yasuo **Arai**, President of Japan Institute of Navigation, Japan,
Prof. Marcin **Barlik**, Warsaw University of Technology, Poland,
Prof. Michael **Barnett**, Southampton Solent University, United Kingdom,
Prof. Eugen **Barsan**, Master Mariner, Constanta Maritime University, Romania,
Prof. Knud **Benedict**, University of Wismar, University of Technology, Business and Design, Germany,
Prof. Tor Einar **Berg**, Norwegian Marine Technology Research Institute, Trondheim, Norway,
Prof. Jarosław **Bosy**, Wroclaw University of Environmental and Life Sciences, Wroclaw, Poland,
Prof. Krzysztof **Chwesiuk**, Maritime University of Szczecin, Poland,
Prof. Daniel **Duda**, Master Mariner, President of Polish Nautological Society, Naval University of Gdynia, Polish Nautological Society, Poland,
RAdm. Dr. Czesław **Dyrcz**, Rector of Polish Naval Academy, Naval Academy, Gdynia, Poland,
Prof. Wiliam **Eisenhardt**, President of the California Maritime Academy, Vallejo, USA,
Prof. Wlodzimierz **Filipowicz**, Master Mariner, Gdynia Maritime University, Poland,
Prof. Börje **Forssell**, Norwegian University of Science and Technology, Trondheim, Norway,
Prof. Masao **Furusho**, Master Mariner, Kobe University, Japan,
Prof. Wieslaw **Galor**, Maritime University of Szczecin, Poland,
Prof. Stanislaw **Gorski**, Master Mariner, Gdynia Maritime University, Poland,
Prof. Andrzej **Grzelakowski**, Gdynia Maritime University, Poland,
Prof. Lucjan **Gucma**, Maritime University of Szczecin, Poland,
Prof. Stanisław **Gucma**, Master Mariner, President of Maritime University of Szczecin, Poland,
Prof. Michal **Holec**, Gdynia Maritime University, Poland,
Prof. Qinyou **Hu**, Shanghai Maritime University, China,
Prof. Marek **Idzior**, Poznan University of Technology, Poland,
Prof. Ales **Janota**, University of Žilina, Slovakia,
Prof. Jacek **Januszewski**, Gdynia Maritime University, Poland,
Prof. Zofia **Jozwiak**, Maritime University of Szczecin, Poland,
Prof. Mirosław **Jurdzinski**, Master Mariner, FNI, Gdynia Maritime University, Poland,
Mr. Adam J. **Kerr**, Editor of the International Hydrographic Review, UK,
Prof. Krzysztof **Kolowrocki**, Gdynia Maritime University, Poland,
Prof. Eugeniusz **Kozaczka**, Polish Acoustical Society, Gdansk University of Technology, Poland,
Prof. Miroslaw **Kozinski**, Gdynia Maritime University, Poland,
Prof. Andrzej **Krolikowski**, Master Mariner, Maritime Office in Gdynia, Poland,
Dr. Dariusz **Lapucha**, Fugro Fugro Chance Inc., Lafayette, Louisiana, United States,
Prof. David **Last**, FIET, FRIN, Royal Institute of Navigation, United Kingdom,
Prof. Joong Woo **Lee**, Korean Institute of Navigation and Port Research, Pusan, Korea,
Prof. Andrzej S. **Lenart**, Gdynia Maritime University, Poland,
Prof. Andrzej **Lewinski**, Radom University of Technology, Poland,
Prof. Józef **Lisowski**, Gdynia Maritime University, Poland,
Prof. Mirosław **Luft**, President of Radom University of Technology, Poland,
Prof. Zbigniew **Lukasik**, Radom University of Technology, Poland,
Prof. Artur **Makar**, Polish Naval Academy, Gdynia, Poland,
Prof. Boyan **Mednikarov**, Nikola Y. Vaptsarov Naval Academy, Varna, Bulgaria,
Prof. Jerzy **Mikulski**, Silesian University of Technology, Katowice, Poland,
Prof. Janusz **Mindykowski**, Gdynia Maritime University, Poland,
Prof. Mykhaylo V. **Miyusov**, Rector of Odesa National Maritime Academy, Odesa, Ukraine,
Prof. Wacław **Morgas**, Polish Naval Academy, Gdynia, Poland,
Prof. Nikitas **Nikitakos**, University of the Aegean, Greece,
Prof. Washington Yotto **Ochieng**, Imperial College London, United Kingdom,
Prof. Stanisław **Oszczak**, FRIN, University of Warmia and Mazury in Olsztyn, Poland,
Mr. David **Patraiko**, MBA, FNI, The Nautical Institute, UK,
Prof. Vytautas **Paulauskas**, Master Marine, Maritime Institute College, Klaipeda University, Lithuania,
Prof. Zbigniew **Pietrzykowski**, Maritime University of Szczecin, Poland,
Prof. Wladysław **Rymarz**, Master Mariner, Gdynia Maritime University, Poland,
Prof. Aydin **Salci**, Istanbul Technical University, Maritime Faculty, ITUMF, Istanbul, Turkey,
Prof. Chaojian **Shi**, Shanghai Maritime University, China,
Prof. Marek **Sitarz**, Silesian University of Technology, Katowice, Poland,
Prof. Wojciech **Ślączka**, Maritime University of Szczecin, Poland,

Contents

Miscellaneous Problems in Maritime Navigation, Transport & Shipping

Introduction

A. Weintrit & T. Neumann
Gdynia Maritime University, Gdynia, Poland

PREFACE

The contents of the book are partitioned into six parts: transportation (covering the chapters 1 through 4), information and computer systems in transport process (covering the chapters 5 through 11), maritime transport policy (covering the chapters 12 through 14), maritime law (covering the chapters 15 through 18), ships monitoring system; a decision support tool (covering the chapters 19 through 25), inland navigation (covering the chapters 26 through 30). Certainly, the subject relating to transport systems and processes may be seen from different perspectives and different branches.

After an introducing to intelligent transportation systems (ITS) and monitoring systems, this book describes the contribution of navigation to land-based traffic management comprising, e.g. in-vehicle navigation systems and advisory routing systems. Besides, an overview of maritime, land and inland traffic management is given.

The first part deals with transportation. The contents of the first part are partitioned into four chapters: The land trans-shipping terminal in processes flow stream individuals intermodal transportion, Modelling of traffic incidents in transport, Maritime transport single windows: issues and prospects and Fire Safety Assessment of Some Oxidizers in Sea Transport

The second part deals with information and computer systems in transport process. The principles of information flow in transport are presented. The contents of the second part are partitioned into seven chapters: Development and standardization of intelligent transport systems, Computer systems aided management in logistics, Information in transport processes, Application of fractional calculus in identification of the measuring system, Railroad level crossing – technical and safety trouble, Application of the Polish active geodetic network for the railway track determination, and The advantage of activating

the role of the EDI - bill of lading and its role to achieve possible fullest.

The third part deals with maritime transport policy. Different approaches to this subject are presented. The contents of the third part are partitioned into three chapters: Effectiveness of the European maritime policy instruments, Sustainable transport planning and development in the EU at the example of the Polish coastal region Pomorskie, and Development of the Latvian maritime policy; a maritime cluster approach.

The fourth part deals with maritime law. The contents of the fourth part are partitioned into four chapters: European Union's stance on the Rotterdam Rules, Maritime law of salvage and adequacy of laws protecting the salvors' interest, The Hong Kong International Convention for safe and environmentally sound management of the recycling of ships Hong Kong 2009, and Maritime delimitation in the Baltic Sea: status iuris.

The fifth part outlines ships monitoring system; a decision support tool. The contents of the fifth part are partitioned into seven chapters: Ships monitoring system, A decision support tool for VTS centers to detect grounding candidates, On the development of an anchor watch supporting system for small merchant ships, Integrated Vessel Traffic Management System for port security in Malaysia, A Simulation Environment for Modelling and Analysis of the Distribution of Shore Observatory Stations - Preliminary Results, The relation with width of fairway and marine traffic flow, and Integrated vessel traffic control system.

The sixth part deals with inland navigation. The contents of the sixth part are partitioned into five chapters: Navigation data transmission in the RIS system, Sea-river technology in transport of energy products, Novel design of inland shipping management information system based on Wireless sensor networks (WSN) and Internet-of-things, Effectiveness of an Integrated Use of Satellite and GIS Technologies on Ships Mixed "River-Sea" Vessels, and

SC-Method of Adaptation Marine Navigational Simulators for Training River Shipmasters.

Transportation

1. The Land Trans-Shipping Terminal In Processes Flow Stream Individuals Intermodal Transportion

A. Kuśmińska-Fijałkowska & Z. Łukasik
Technical University of Radom, Radom, Poland

ABSTRACT: The influence has the system of the exchange on enlargement of the efficiency of the work trans-shipping land terminal and processing information, which in he will improve the processes of the flow of the individuals of the intermodal transportation which was presented in the figure of algorithms more considerably to the measure. The algorithmization makes possible execution mathematical analyses, as also the influence has on the quality of the executed processes of the flow of the stream of the individuals of intermodal transportation in the trans-shipping land terminal.

1 INTRODUCTION

Correct functioning the whole chain of intermodal transportation depends in the considerable measure on proper functioning land terminals, and in this first of all from their ability infrastructural to the executing the trans-shipments, cost, range of offered services, quality and reliability. Presented algorithms in the article, in the considerable stage will contribute to improvement of the processes of the flow individuals in the land terminal to prevent in the future the situation from Fig. 4 obviously they were created stay on the basis of the observation of the real object in which shortcomings were observed. The algorithmization makes possible in so folded system what the land terminal is to conduct the effective mathematical analysis of drawing ahead processes. Because the modern terminal of intermodal transportation is more the than simple trans-shipping point and develops in the direction of creating the centres of the service of the transport of cargos about the wide range of offered services. (Łukasik Z., Kuśmińska A., Matejek T. 2006-2007)

2 ANALYSIS OF THE STATISTICAL DATA ARRIVING THE INDIVIDUALS INTERMODAL TRANSPORTATION TO LAND TERMINAL ON TRAIN

Road vehicles arrive on land terminal (Fig. 1) to deliver the individuals on given string of cars, the larger number arrives with the considerable store of the time before string of cars. These attentions are confirmed by conducted investigations. The switching

of the branch of road and railway transportation is the aim of applying intermodal transportation in net and their integration within the of the general conception. (Kuśmińska A., Łukasik Z. 2005)

Figure 1. The individuals on entry and exit land terminal (Kuśmińska A., Łukasik Z 2005)

The presented option of arriving the individuals on the land terminal road vehicles (Fig. 2), the arriving the individuals on the terminal represents, a lot of earlier the before string of cars he is prepared to trans-shipping operations, and now he the trans-shipping device has to execute double operations in this case with, what operating costs, the time of the expectation of individuals join obviously.

The solution eliminating the indirect trans-shipment represents the Fig. 3 individual they arrive in such spaces of the time that they are subjected the direct trans-shipment from the road vehicle on the wagon of string of cars.

It was affirmed in the result of the analysis of statistical data gathered from the observation of the real object that individuals arrived the road vehicle to land terminal with considerable temporary superiority on given string of cars. This introduced on the graph in the co-ordinate time became before the string of cars will lower the terminal, and day of arrival of individual (Fig. 4). What the situation from the real object is the very „poor" case he answers the option from graph (Fig. 2).

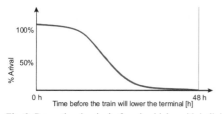

Fig. 2. Proportional arrival of road vehicles with individuals in the function of time (poor option of the arrival) (S.C.2.70 Deliverable 3. 1999)

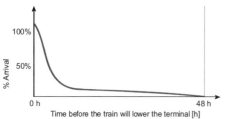

Fig. 3. Proportional arrival of road vehicles with individuals in the function of the time (best option of the arrival) (S.C.2.70 Deliverable 3. 1999)

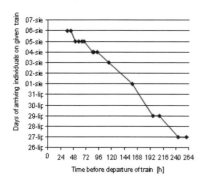

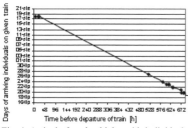

Fig. 4. Arrival of road vehicles with individuals in the function of the time [h] (own study)

Intermodal transportation gained temporary superiority over road transportation in our national conditions, you should improve the processes of arriving the individuals road vehicles to trans-shipping terminal what is visible on the graph in co-ordinates % arriving individuals the road vehicle to land terminal, and the time before the string of cars will lower terminal (Fig. 4), and also improve the flow of individuals in the terminal itself which is the bonding

bar of the road and railway piece. (Kuśmińska A., Łukasik Z. 2006)

3 ALGORITHMS OF THE REALIZATION PROCESSES IN THE TRANS-SHIPPING LAND TERMINAL

The influence also has efficiency of realization of processes such on effective functioning land terminal (Fig. 1) how:
– analysis offer ask;
– realization of orders;
– railway- road service;
– road-railway service.

The presented algorithm of functioning land terminal (Fig. 5) will make possible to improve processes drawing ahead in the land terminal, how also the chain of intermodal transportation.

Analysis Offer asks (Fig. 6) this the first socket with the customer the party is whose aim as the largest number of orders, the question offer from the customer is delivered in the written mould on the terminal. Directed to the Trade Aggregate the aim of preparing the offer becomes after executing the registration, the party is the priority obviously as the largest number of orders. The Trade Aggregate should check in first order, if he is in the state match the customer during realization to requirements the order, if you should so execute all steps havings on the aim of the receipt of given to the correct calculation costs and prepare the offer. Then after checking and identify, that the offer was prepared well sent to the customer (fax, e-mail) becomes. After dispatch offers, responsible worker for the correct process of offering the services, he contacts with the customer the aim of making sure, what to the regularity of conditions and execution of the interview, what to possible her party. In the case of the settlement of the incompatibility of the conditions of offer from the customer expectation the worker after the consultation with superior makes the corrections of the offer. However, while the introduced offer becomes accepted by the customer the next stage of the process is realized - **the realization of order** (Fig. 7).

The realization of the order which the realization of services is the main aim: railway - road, road- railway in as the shortest time (Fig. 7) the order of the realization of the service is delivering to trans-shipping terminal in the written mould (fax, e-mail). The realization of the order follows after executing the analysis of the agreement of working plan and the identification of the realization of the service. While executing the order the opinion of the regularity of the realization is made. Responsible aggregate for the realization of the order represents the conclusion to the management terminal in the case of incompatibility from the customer expectation. When the customer undertakes reclamation workings begun then becomes the peaceable conduct with the internal instruction.

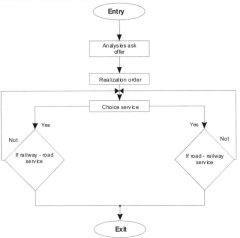

Fig. 5. Algorithm of functioning land terminal (own study)

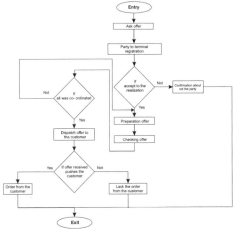

Fig. 6. Algorithm of the analysis offer asks (own study).

During the realization of the order you should haul the kind of the realization of the service:
– railway - road,
– road - railway.

The main aim of the service is the transport of intermodal individuals on near the minimization of the standstill of individuals on the terminal customer order **railway - road** (Fig. 8). While individuals on wagons are on the trans-shipping track he follows identification, agreements with the transports letter

(working plan) (technology RFID). The dispatcher prepares them to the landing after affirming agreement and the party of documents (letter transporting-identifying individuals). Individuals subjected the direct trans-shipment become unblocked on wagons.

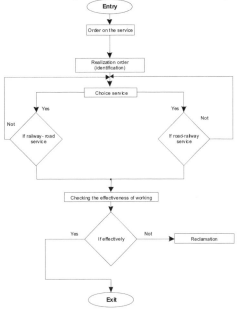

Fig. 7. Algorithm of the realization of orders (own study).

The choice of the technology of the trans-shipment follows in dependence from: the condition of the individual: or vain, if capacious, the type: container 20', replaceable body etc. and also her technical state. He next is considered the decision, which to use the device during the trans-shipment, if the gantry, if the jacks cart, and to accelerate two simultaneously maybe trans-shipping works. Already possessing information the operator of the device places them if the given individual is subjected the direct trans-shipment on the fix to the trans-shipment after executing the choice i.e. he can be charged on the waiting on her wagon or on the road vehicle. He follows the identification of wagon or road vehicle and direct trans-shipment, the protection of the individual and documentary evidence the realization of the service.

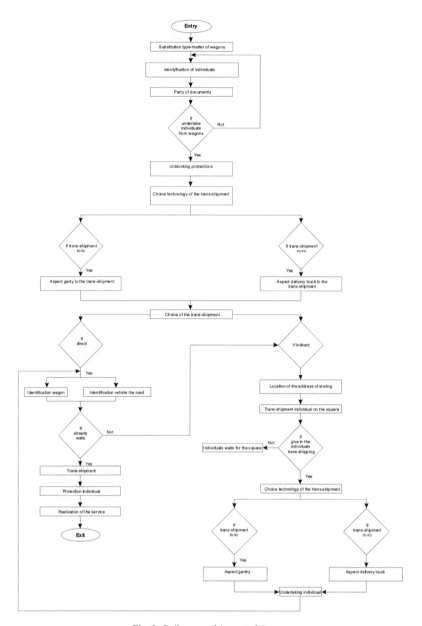

Fig. 8. Railway-road (own study).

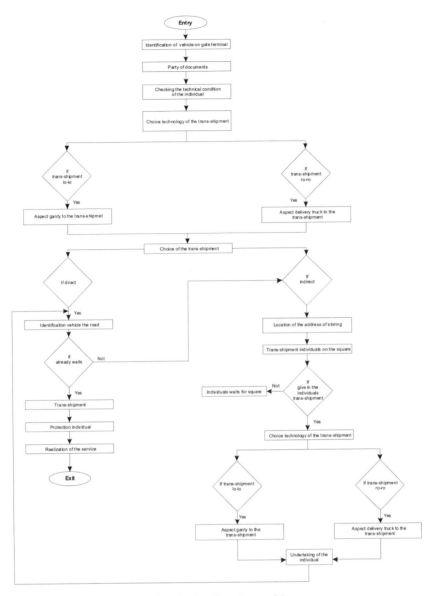

Fig. 9. Road - railway (own study).

In the case, when road vehicles did not arrive on the time of the individual subjected the indirect trans-shipment find, and now the operator of the trans-shipping device before he will undertake the individual from the wagon he knows her exact location on the component square. Trans-shipping works begin then and putting individual in sectors on the component square. When the delayed road vehicle arrives on the land terminal after given individual her identification (technology RFID) follows on the square then, the choice of the technology of the trans-shipment, trans-shipment, the protection of in-

dividual on the semitrailer of the road vehicle and last stage documentary evidence the realization of the service. The land terminal leaves after executing all actions individual.

The algorithm of **the road-railway service** the realization of the operation of the efficient trans-shipment whose aim minimalizing the time is also spent individuals on the land terminal was introduced on (Fig. 9) While the individuals arrive on the land terminal the road vehicle on the gate subjected the identification, the type of individual, protection, technical state become, received documents become then and the decisions are made on the basis of contained information, what to the choice of the technology of the trans-shipment (the trans-shipment lo-lo perpendicular, if horizontal ro-ro like simultaneously lo-lo, ro-ro the aim of acceleration of the works of trans-shipping). The operator prepares them to the trans-shipment after the choice of the device. In the case, when he undertakes the individual from the semitrailer of the road vehicle he possesses information, if the given individual is subjected trans-shipping direct, if indirect. The operations of the direct trans-shipment are the priority. After identification the wagon on, which the individual has to settle the operations of the trans-shipment begin protection her and documentary evidence the realization of the service then. In the case of the indirect trans-shipment the operator before he will undertake the individual from the semitrailer of the road vehicle he knows exact location and the sector of assembly her on the square (RFID) because the wagons are not prepared to the trans-shipment yet. When wagons be prepared, one can begin the works loading of the individuals which are subjected identification on the square terminal (RFID). The next stage is the choice of the technology of the trans-shipment and undertaking of the individual from square and placing her on the wagon after identification individual. When all individuals are already on wagons moulding the string of cars safe become, in the documents of the realization of the service are prepared between the time. The type - matter of wagons with individuals (conferment) leaves the land terminal after executing all formalities.

4 CONCLUSIONS

The integration of forwarding processes, he requires the development of international systems joint in one net about the large possibilities of the flow of the freight pulp. One talks about the intermodal transportation which unites railway transports with road and sea near one figure of the cargo behaviour more and more often. Support of the development of terminals in the chains of intermodal transportation, the limitation of the transports of goods will let suppose the road transportation which guides not only to decay of the road infrastructure but enlargement of the release of exhaust gas also causes what he guides to the degradation of the natural environment in the consequence. Strong exists so one need the set - back of the tendency to growing utilization of car transportation on the thing of intermodal transportation.

One notes down on the Polish forwarding market of the hesitation of the pace of the growth since several years and the lack of the uniform and dynamic development of transports intermodal. He is the result of this the worsening quality of railway services, the growth of the time of the service of intermodal individuals on terminals (among others problems with identification individuals on the component square), the deepening price uncompetitive of railway transportation in the relation to car transportation and the lack of the complex and effective instruments of the forwarding politics of the state, supporting the intermodal transportation. According to the expectations of the growth of intermodal transportation in the Poland, the terminals have to be prepared on the party of the larger number of inflowing streams individuals. (Kuśmińska A., Łukasik Z. 2005) many of Polish land terminals the present moment stays trans-shipping terminals from the name exclusively. Land terminals functioning on the terrain of our country in the future matched charged tasks should also improve the system the processing and the circulation the information, which together with the system of steering the stream individuals, he was presented in the figure of algorithms and which he will improve the processes in the land terminal.

LITERATURE

Kuśmińska A., Łukasik Z.: „Models trans-shipping processes in the intermodal terminal" Technical University of Radom, „LogiTrans", Szczyrk 2005

Kuśmińska A., Łukasik Z.: „Models the processes of the flow JTI in the land terminal", „IntLog" Stockholm 2006.

Łukasik Z., Kuśmińska A., Matejek T- (stage II),: „ Intermodal transportation in the processes of the flow of cargos" the stage II, III, Technical University of Radom 2006-2007

S.C.2.70 Deliverable 3: the „Design of of Platforms Simulation environmentt" Lugano 31.01.99

2. Maritime Transport Single Windows: Issues and Prospects

K. E. Fjortoft, M. Hagaseth
MARINTEK-SINTEF, Norwegian Marine Technology Research Institute

M. A. Lambrou
University of the Aegean, Department of Shipping, Trade and Transport, Chios, Greece

P. Baltzersen
Wilhelmsen IT Services, Norway

ABSTRACT: In the trade, transport and shipping sector, the Single Window (SW) concept has been evolved over time in a number of forms, reflecting respective policy, regulatory, market and technological regimes of the domain. A SW primarily addresses the need for efficient electronic transactions between governmental and business entities; however the SW service model adopted by the responsible authority and the offered SW system functionality differ; currently at least two distinct approaches are observed, namely a customs-centric SW approach, and a maritime and port centric approach. In all respective cases, the SW service model, the SW ownership model (public, private or Private-Public-Partnership), legal and regulatory aspects and the SW revenue model (free or with a fee) consist pertinent SW service design issues. Thus, different types of SW systems evolve in terms of offered service bundle, namely ship clearance, cargo import/export, or port clearance SWs, where often vested interests and policy choices dictate the dominance of one model implementation over the other. Modern ICT tools may significantly help to organize and improve the efficiency of a SW design and implementation process. In this paper, admissible development frameworks and methodologies are examined towards the efficient implementation of SW service models that are explained. Our analysis is based on experiences gained in the Norwegian SW national initiative (http://www.sintef.no/Projectweb/MIS/) and the EU eFreight project (http://www.efreightproject.eu/).

1 INTRODUCTION

The one stop shop business model has been exhaustively researched and applied in the context of e-business and e-government service provision over the last decade (Wimmer, 2002; Lambrou et al., 2008). In a similar vein, in the trade, transport and shipping sector, the "Single Window" (SW) concept was formalized by the United Nations Centre for Trade Facilitation and Electronic Business (UN/CEFACT 2005) to enhance the efficient exchange of information between trade and government agencies. The Single Window concept has its origin in the Trade Facilitation and Customs field focusing upon efficient import and export institutions and mechanisms, where declarations of goods related to regulatory information must be reported in cross border activities.

A SW primarily addresses the need for efficient and collaborative electronic transactions between governmental and business entities; however the co-coordinating SW authority and the core functionality may differ, thus we typically observe a customs-centric, import and export oriented approach, a port and ship oriented (maritime focus), and a safety and security centric approach. In both cases pertinent

SW service design aspects include the SW ownership model (public, private or public-private partnership - PPP), and the SW cost model (e.g., free use, membership or transaction fee). The organizational level of the SW competent authority, e.g., international, national, regional, or local is an important differentiating factor, as well. Often, vested interests and policy choices dictate the dominance of one model implementation over the other.

In this paper we discuss different types of single window systems and enabling development methodologies and platforms, focusing in particular on a maritime centric model where ship clearance, cargo import/export, and port clearance services are supported. This means that we extend the Single Window Concept to conver not only regulatory information, but also other information related to maritime transport.

2 A TAXONOMY OF SINGLE WINDOW SYSTEMS

There are different reference models of existing and emerging SW systems supporting intermodal transport activities, as explained in Table 1. A SW system can cover the cargo reporting activities where import and export declarations are the main processes supported, another SW model is organized around ship or vessel clearance activities offered by national governments, whereas a third model is a port clearance oriented SW. The purpose of a ship oriented SW is to support all mandatory information reporting concerning a ship sailing from abroad to a EU or associated country, as based on the SafeSeaNet (SSN) system notifications and formalities.

All countries in EU and Associated countries are connected or will soon be connected to the central SSN system. Every country has to dedicate an internal authority as a National Competent Authority that will be the official connection between the country and the central SSN system that is under the responsibility of the European Maritime Safety Agency, EMSA.

A Port Single Window (PSW) can in many cases be defined as a Port Community System (PCS). It is a community system which based on an integrated series of procedures, rules, standards and ICT solutions supports the automatic exchange of data and documents related to the port authorities' clearance of ships and cargo upon arrival, stay and departure of vessels.

A PSW is primarily supporting the requirements of governmental agencies, but also the requirements of the cargo parties' interests. So a PSW covers Customs requirements and document handling, and the information exchange dealing with the necessary services in a port and the handling of ship and cargo. It is also likely that a PSW will have a stronger focus upon private information and more commercial oriented regarding sale and ordering of port services than the one for ship clearance.

EPC (Electronic Port Clearance) is the concept used to refer to vessels visiting a port and their electronically (without the use of paper documents) dealing with all formalities, documentary requirements and procedures associated with the arrival, stay and departure of ships engaged on international voyages. On the one hand, EPC aims to replace the paper documents such as the FAL Forms currently in use; on the other hand EPC tries to make the exchange of information more efficient, through the rationalization of the procedures and simplifying the related data. Figure 1 gives an overview of the three dimensions of Single Window systems and how each of them relates to each of the actors. Note that the actors *Ship Owner* and *Charterer* only interacts with the Single Window through other systems, not as separate actors. The actor *Other Port Parties/ 3rd Party Systems* includes parties involved in the port business other than the port authorities, for instance systems to handle resource bookings.

Table 1. Types of Single Window

Single Window for cargo	
Description	A SW for customs clearance normally contains information about cargo for either import or export.
Users	The users are Consignor's and Consignee's, the Customs, as well as cargo agents
Characteristics	The goods to be defined for import and export will need a release number before the transport can progress from an import area at a terminal. A main functionality for this SW is the cargo clearance process.
Objects	Cargo information and definition, Ownership, The itinerary of the goods, Handling instructions, General statusinformation about the cargo
Functionality	Registers: Goods group, Location register, Tax code Automation: XML and Web-based user interface Accessibility control Hand-over mechanism with other SW-solutions

Single Window for ship clearance	
Description	A SW for Ship clearance contains information about the ship, the voyage, the cargo, the passengers, the crew and information that is required by the SafeSeaNet directive.
Users	The users are the ship it self, agents, the providers on the ship, or the governmental bodies that need statuses and information for controlling duties, for mainly safety/security purposes. Governmental bodies can be Police, Coast Guard, Navy, Coastal Administration, Health authorities, or the ports.
Characteristics	The main purpose of such a SW is to have a good overview of the safety and security issues regarding sea transport. It could be either a site where information about a ship transport could be achieved in a distressed situation, or it could be more used in a controlling purpose where i.e. the crew and passenger list is matched with the list of criminals by the Police authorities. A main functionality for this SW is the ship clearance process.
Objects	Ship information, Cargo information, Crew and Passenger information (also effects), voyage information. Notification messages (hazmat, security, alert, ship) between the different states should also be considered
Functionality	Registers: Goods group, Vessel, Location Automation: XML and Web-based user interface (both ways) Acceptance report/Clearance notification (automatic) Use of sensor data for report purpose Ordering of transport services such as pilot age services Hand-over mechanism with other SW-solutions as well as commercial systems from service providers

Single Window for port clearance	
Description	A SW for port clearance is a reporting site for needed information regarding an entrance to a port. The information could also be about information classified as private, and used within a commercial aspect.
Users	The ship, the ship operators, the agents, the port management, the port service providers
Characteristics	This SW is used to achieve a port clearance of a ship. The information is both of a private and a public character. The ports are using the information to plan the ship entrance, to achieve the port safety and security regulations, and to calculate the fees to be sent to the users. A main functionality for this SW is the port clearance process.
Objects	Ship, Cargo, Load units, Service needs, Security information
Functionality	Registers: Goods group, Vessel, Location, Port services. XML and Web-based user interface (both ways). Acceptance report/Clearance notification (automatic), Use of sensor data for report purpose, Safety and security, Ordering of port services, Accessibility control, Hand-over mechanism/communication mechanisms with other SW-solutions as well as commercial systems from service providers, Statistics, General port information, Site for laws and regulations.

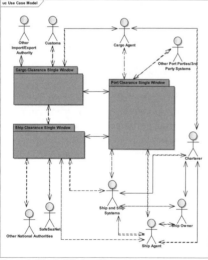

Figure 1. Single Window Taxonomy

One of the challenges in the specification of SW system is to decide the dimensions and geographical areas the SW system should cover.

Examples of such dimensions are:
- International dimension
- National dimension
- Regional dimension
- Local dimension

Another dimension could be an Ad-hoc solution.

For all those dimensions there are different needs and a different legal basis to follow that also differs within one dimension. An example can be that within a port, which is defined as a local dimension solution, there are some port specific regulations to follow regarding mandatory reporting and the configuration of the SW must therefore follow the properties defined at the port where also the private-public partnership relations must be placed, Figure 2. This means that a Single Window system for one port may differ in several respects to a Single Window system in an adjacent port.

It is likely that the different systems that represent the different solutions must exchange information with each other. A ship normally crossing the defined dimensions where coming from an abroad country and visiting a port, is mirroring the process when returning to an abroad port. In such a case, the ship must first follow the local dimension from departure port, where regulations and other procedures are followed (reporting time, place, etc). Then, in some cases, new reporting and procedures must be followed when sailing in a certain region such as a fjord or vulnerable regional areas. Then, when the ship is leaving the national waters, information must be reported to the NCA (National Coastal Authorities) or the Coast Guard of the departure nationality, and finally follows international conventions in open international waters. The same approach will be relevant when sailing into the arriving port.

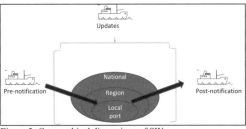

Figure 2. Geographical dimensions of SW

3 DEVELOPMENT METHODOLOGIES FOR SINGLE WINDOW SYSTEMS

Several methodologies relevant for the development of Single Window systems can be exploited. A number of available methodologies focus on the analysis and design phase using various process modeling techniques, while other methodologies are related to the technical implementation of a SW system (e.g. SoaML).

3.1 Zachman Framework

The Zachman framework (Zachman, 1997) was first presented in 1987 and has since then evolved in several directions and several versions. For maritime Single Window development, it is most relevant to view it as a taxonomy for organizing architectural artifacts, design documents, specifications and models. The framework addresses the question of who is the target for the description and also what is described, for instance data and functionality. In this sense, the Zachman Framework is not a methodology since it lacks methods and processes for collecting the information, and also for managing or using the information. Rather, Zachman describes the framework for enterprise architecture as follows: *"The Framework as it applies to Enterprises is simply a logical structure for classifying and organizing the descriptive representations of an Enterprise that are significant to the management of the Enterprise as well as to the development of the Enterprise's systems."* A key point in the Zachman framework is that the same complex item can be described for different purposes in different ways using different types of descriptions.

The framework has 36 categories for completely describing anything related to the enterprise, organized with six columns and six rows. Each row represents a total, distinct and unique view of the solution from a particular perspective. Each column represents a category of the enterprise architecture component, called focus. These are data description (what), function description (how), network description (where), people description (who), time description (when), and motivation description (why).

Some aspects of the Zachman framework that are convenient for analyzing SW systems include:
1 Analysis of several organizations that have to cooperate in an interoperable SW system:
A SW is an environment which has to support interoperability among highly heterogeneous environments. This means that a structured way to present the analysis of the organizations with different viewpoints is important. The Zachman Framework for systematically describing changes to an organization based on various viewpoints and various abstraction levels is very useful in the analysis phase of a SW development.
2 Clarification of different views of the same artifact:
The Zachman Framework focuses on different views of the same artifact (process, data), which is important in a SW system covering processes and data originating from various applications, both cargo, port, and ship clearance, but also orig-

inating from both public and private organizations.

3 Presentation of analysis results throughout several organizations:

The Zachman Framework can be useful to present the analysis of a SW system. Important here is the fact that new third party systems that want to collaborate with the SW may have easier access to the taxonomy of the SW.

Some aspects of the Zachman framework that are missing for SW systems:

1 Lack of a structured methodology:

The Zachman Framework does not include a methodology per se, however, a methodology is needed for the Single Window design and implementation process, for instance for describing how to integrate a third party system with the Single Window, and how to handle and share data that is specific for the third party systems.

2 Description of both the SW and the third party systems are needed:

The Zachman framework seems to focus on the description of a single enterprise, however, when designing a SW system, we have to consider several organizations and environments as a whole. This is because each service provider and service user may represent distinct organizations with their own Zachman matrix related to Single Window. What is needed, is a Zachman Framwork analysis of the Single Window itself, but in addition, we would need to have descriptions of the third party organizations, at least the parts that are most relevant for the Single Window system.

3.2 CIMOSA

CIMOSA (Computer Integrated Manufacturing Open System Architecture) is an enterprise modeling framework, which aims to support the enterprise integration of machines, computers and people. The framework is based on the system life cycle concept, and offers a modeling language, methodology and supporting technology to support these goals. Three dimensions of CIMOSA are outlined (Zuesongdham, 2009):

1 The generic dimension (Instantiation of Building Blocks) is concerned with the degree of particularisation. This dimension differentiates between Reference Architecture and Particular Architecture.

Reference Architecture resembles a catalogue of reusable building blocks which contains generic and partial building blocks applicable to specific needs.

Particular Architecture serves the use of a specific case in process modelling which is not intended to be reusable for other models.

2 The modelling dimension (Derivation of Models) provides the modelling support for the system or

work life cycle starting from requirements to implementation.

3 The view dimension (Generation of Views) offers the users to work with partial models representing different aspects of the enterprises: function, information, resource and organisation with the option for other views to be defined as needed.

Advantages of CIMOSA:

1 Strong inter-organizational process modeling:

The CIMOSA is strong on modeling complex organizations and has constructions to model different views of the same things in an organization.

2 Use of reference architecture:

The reference architecture can be used to build up an library of Single Window concepts which may be useful when new third party systems are to connect to a Single Window. Then, reuse of common descriptions and concepts may be facilitated through the use of CIMOSA.

3 Focus on common understanding of terms:

CIMOSA has focus on defining a glossary for common understanding of terms and definitions.

Problems with using CIMOSA:

1 Process descriptions separate from implementation aspects:

CIMOSA does not cover the technical implementation phase, for instance related to ICT architecture or software services. However, attempts have been made to extend this framework (Zuesongdham, 2009)

2 Complex framework:

The framework may look complex, since it has a three dimensional matrices describing the modeling framework.

3.3 SOA and SoaML

SOA (Service Oriented Architecture) is an architectural paradigm whose goal is to achieve loose coupling among a collection of interacting software services. Services are usually defined as autonomous, platform-independent computational elements that can be described, published, discovered, composed and consumed using standard protocols for the purpose of building distributed, collaborative applications within and across organizational boundaries (Manolescu et al., 2005, Rødseth et al., 2011).

SoaML (Service oriented architecture Modeling Language) is an open source specification project from the Object Management Group (OMG), describing a UML profile and metamodel for the modeling and design of services within a service-oriented architecture (Casanave, 2009). SoaML meets the mandatory requirements of the UPMS (UML Profile and Metamodel for Services). SoaML includes descriptions of how to identify services, the requirements they are intended to fulfill, the functional capabilities they provide, what capabilities consumers are expected to provide, the protocols or rules for using them, and the anticipated dependencies between them.

Advantages of using SoaML for Single Window Systems:

1 Closer connection between the design and the implementation phases of the Single Window development:

SoaML ensures a close connection between the description and the implementation of services, since SoaML works at the level of services. This means that the connection between for instance web services to implement the services can be very tight to the description of the service architecture of the Single Window system.

2 Focus on services both at the design and implementation level:

Using SoaML means that both analysis and implementation will focus on services. This is important in at least two aspects: to avoid silo thinking regarding the systems, and to facilitate the creation of complex services based on basic, already existing services in the Single Window environment.

3 Intuitive way to describe third party systems:

This is important in the sense that the third party systems can be described as legacy systems being wrapped up in a new service. The third party systems can be described with a clear interface to the Single Window system regarding information exchange, functionality and payment regimes.

4 Reuse of services in new, complex services:

A SW system will make more information than before available as a whole. This means that the information can be combined in new ways offering new services. This leads to the need to combine new and existing services into complex services. SoaML is suitable for describing such complex services.

Difficulties of using SoaML for SW Systems

1 Unclear notion of data model since the focus is on services and processes:

The semantics of data in a Single Window environment is very important since several systems with heterogeneous data models have to cooperate. In this context, the notion of *three level architecture* for data model is important, that is, the *conceptual schema* describing all concepts (entities and relationships) in the Single Window, and the *external schema* describing only the part of the data that is relevant for each third party system. Separate from this, we have the *internal schema* describing the implementation. SoaML does not contain explicit description of this.

2 Complex syntax:

The notation and syntax of SoaML may appear to be complex for those who are not familiar with it. This may lead to difficulties in the communica-

tion with practitioners in the port and logistics domain.

In (Zuesongdham, 2009), it is argued that a combination of CIMOSA and SOA modeling is the most attractive approach to develop interoperable port community systems. This is because the inter-organizational process modeling capabilities from CIMOSA is needed in addition to the more technological related implementation view.

4 CONCLUSIONS

Several definitions of Single Window exists, most of them are focused on handling regulatory information. The future will show a more integrated information society between public and private systems as this paper has shown. However, we have argued that the taxonomy and the use of commonalities between them must be in place to have a harmonized way of doing trade and transport. If each of them will be developed as a proprietary system than lots of engineering between them is needed to get any valuable benefits of using the Single Window concepts as described. One of the key points is use of standard information and code values.

Several initiatives related to the development and operation of maritime Single Window systems exist, including the description of Single Window frameworks by IMO (IMO/FAL36/1 2010, IMO/FAL/36/2 2010), UN/CEFACT recommendation on Single Window (UNCEFACT 2005, 2010), and (APEC 2009), and the description of several data models related to Single Window including (ISO/FDIS 28005-2 2010) on Electronic Port Clearance (EPC), and (ISO 7372 2005) on TDED data model. However, it is important to further proceed in obtaining a distinct and unified framework and methodology for developing Single Window systems, which will crucially support a smooth and manageable integration of heterogeneous systems into a Single Window environment. Single Window systems need to be developed based on the compatible standards regarding formal descriptiona of logistics processes, interfaces and information content. Cooperation between several national and regional Single Window solutions will be simplified if the systems are developed based on a unified framework and compatible methodologies. Also, integration of the numerous third party systems into the different national Single Window environments will be more efficient as based on a unified methodological background, and if systematically applied.

ACKNOWLEDGEMENTS

This work has been partially funded by the EU eFREIGHT 7FP DGTREN Project (www.efreightproject.eu) and by Norwegian Research Council MIS project (Maritime Information Centre- http://www.sintef.no/Projectweb/MIS/).

REFERENCES

Casanave, C. 2009. Enterprise Service Oriented Architecture Using the OMG SoaML Standard, white paper on www.modeldriven.com

CIMOSA home page: http://www.cimosa.de/

IMO/FAL/36/1 2010. Electronic Means for the Clearance of Ships: The use of the Single Window Concept, Submitted by Brazil.

IMO/FAL/36/2 2010. Electronic Means for the Clearance of Ships: Justification for Single Window Guideline for Maritime Transport, Submitted by the Republic of Korea.

ISO/FDIS 28005-2 2010. Security management systems for the supply chain -- Electronic port clearance (EPC) -- Part 2: Core data elements

ISO 7372 2005. Trade data interchange -- Trade data elements directory

Lambrou, M. A., Pallis, A. A. and Nikitakos, N. V. 2008. 'Exploring the applicability of electronic markets to port governance'. International Journal of Ocean Systems Management, 1, 14-30.

Rødseth, Ø.J., Fjørtoft, K, Lambrou, M 2011. Web Technologies for Maritime Single Windows, Proceedings of MTEC 2011.

UN/CEFACT 2005. Recommendation No.33: "Recommendation and Guidelines on Establishing a Single Window".

UN/CEFACT 2010. Recommendation no 35: Establishing a Legal Framework for International Trade Single Window.

Wimmer, M.A. 2002. 'Integrated service modeling for online one-stop Government', Electronic Markets, Vol. 12, No. 3, pp.1–8.

Zachman, John: Concepts of the Framework for Enterprise Architecture, 1997.

Zuesongdham, P. 2009. Combined Approach for Enterprise Modeling: CIMOSA and SOA as Dynamic Architecture in Maritime Logistics.

3. Modelling of Traffic Incidents in Transport

J. Skorupski

Warsaw University of Technology, Faculty of Transport, Warsaw, Poland

ABSTRACT: Safety is one of the most important criteria for assessing the transport process. The traffic process in available traffic space are partly organized and planned. However, these plans are subject to numerous disturbances of probabilistic nature. These disturbances, contribute to the commission of errors by the operators of vehicles and traffic managers. They lead to traffic incidents, which under certain circumstances may transform into accidents.

In the paper the method of modelling traffic incidents, using different types of Petri nets is presented. Example of the serious air traffic incident shows the opportunities offered by the application of this modelling technique. In addition, the possibility of its use in maritime transport, for example, modelling of traffic at the waterways intersection is presented.

1 INTRODUCTION

Transport is a complex system combining advanced technical systems, operators and procedures. All these elements work in a large spatial dispersion, but are closely interrelated. They interact, and the time horizon of these interactions is very short. In sea, air or railway transport, the risk is traditionally identified with the accidents, which typically produce a high number of deaths and huge financial losses. Severity of the consequences is the reason why the safety was always a key value in transport.

For instance Polish aviation regulations define three categories of events (Aviation Law, 2002):

- accident - as an event associated with the operation of the aircraft, which occurred in the presence of people on board, during which any person has suffered at least of serious injuries or aircraft was damaged,
- serious incident - as an incident whose circumstances indicate that there was almost an accident (such as a significant violation of the separation between aircraft, without the control of the situation both by the pilot of the aircraft and the controller),
- incident - as an event associated with the operation of an aircraft other than an accident, which would adversely affect the safety of operation (e.g. a violation of separation, but with the control of the situation).

In this paper, traffic incidents in transport are subject of interest. A method for modelling these incidents with use of Petri nets theory is presented. This method allows the analysis of the causes of incidents as well as assessing the probability of transformation of incidents into accidents. The method uses coloured, stochastic, timed Petri nets, with the time assigned to markers.

The first part of the paper presents basic information about Petri nets. The next discusses the specificity of the analyzed transport systems and a method of modelling those using coloured, timed Petri nets. The next two chapters contain examples of analysis using the proposed method. The first example comes from the air traffic and presents the calculation of the possibility of transforming a serious incident into an accident (Skorupski 2010). The second example concerns the maritime traffic and demonstrates the applicability of the method for modelling conflict at the intersection of the waterways.

2 THE BASICS OF PETRI NETS

Petri nets provide a convenient way to describe many types of systems. Especially a lot of applications they found in software engineering, where they are used particularly to describe and analyze concurrent systems. There is a rich literature in this subject, e.g. (Jensen, 1997, Szpyrka, 2008), which also contains an extensive bibliography of the topic. In this paper it was shown that Petri nets can also be used for modelling transport systems, particularly the traffic processes. The examples concern the analysis of traffic safety problems in air and maritime transport, but a similar approach can be applied to other modes of transport.

2.1 Types of Petri Nets

Depending on the needs, one can define different Petri nets with certain properties. However, there is a set of characteristics that are common to such networks. The basis for building a Petri net is a bipartite graph containing two disjoint sets of vertices called places and transitions. Arcs in this graph are directed and single, and therefore it is a Berge graph. A characteristic feature of the graph used in Petri nets is that the arcs have to combine different types of vertices. Below are presented brief definitions of basic types of Petri nets: first low, then a high level (Marsan et al. 1999). Detailed analysis of the properties of various types of nets is included in the literature and will not be discussed here.

2.2 Generalised Petri net

Generalised Petri net (GPN) is described as:

$$N = \{P, T, I, O, H\} \tag{1}$$

where:
P - set of places,
T - set of transitions, $T \cap P = \phi$,
I, O, H, are functions respectively of input, output
      and inhibitors:
$I, O, H: T \rightarrow B(P)$
where $B(P)$ is the superset over the set P.

Given a transition $t \in T$ it can be defined:
$t^+ = \{p \in P : I(i, p) > 0\}$ - input set of transition t
$t^- = \{p \in P : O(i, p) > 0\}$ - output set of transition t
$t^o = \{p \in P : H(i, p) > 0\}$ - inhibition set of transition t
GPN is characterized by the fact that the functions described on arcs: $I(t,p)$, $O(t,p)$ and $H(t,p)$, can take values greater than 1, which is equivalent to the presence of multiple arcs between nodes.

2.3 Marked Petri net

Marked Petri net (MPN) is described as:

$$S_M = \{P, T, I, O, H, M_0\} \tag{2}$$

where: $N = \{P, T, I, O, H\}$ - generalised Petri net,
$M_0 : P \rightarrow \mathbf{Z}_+$ is the initial marking, i.e. a function assigning an integer to each place.

We also say that the marking specifies the number of markers assigned to each of the places.

Initial marking, along with the rules governing the dynamics of the net, that is rules of marking changes, determine all possible reachable markings.

The same network but with different initial markings will describe different systems.

Transition t is called active in marking M if and only if:

$$\forall p \in t^+, M(p) \geq I(t, p) \wedge \forall p \in t^o, M(p) < H(t, p) \tag{3}$$

Firing of transition t, active in marking M removes from any place p belonging to the set t^+, as many markers as function $I(t,p)$, determines. At the same time it adds to any place p from the set t^-, as many markers as determined by the $O(t,p)$, function. This means firing of transition t will change actual marking to M' such that

$$M' = M + O(t) - I(t) \tag{4}$$

This relationship is written briefly $M[T\rangle M'$. We then say that M' is reachable directly from M. If the $M \rightarrow M'$ transformation requires firing a sequence of transitions σ, then we say that M' is reachable from M and denote $M[\delta\rangle M'$.

2.4 Place-transition Petri net

Place-transition net (PTN) is a generalized, marked Petri net, supplemented by the characteristics of places interpreted as their capacity, i.e. the maximum number of markers that can accommodate any of the places. Thus, a place-transition net can be written as

$$S_{PT} = \{P, T, I, O, H, K, M_0\} \tag{5}$$

where: $N = \{P, T, I, O, H\}$ - generalised Petri net, $K : P \rightarrow \mathbf{N} \cup \{\infty\}$ – capacity of places, and the symbol ∞ means that a place has unlimited capacity, $M_0 : P \rightarrow Z_+ \wedge \forall p \in P : M_0(p) \leq K(p)$ – initial marking.

2.5 Timed Petri net

With timed Petri net (TPN) we have to do, when firing a transition is not immediate, but it takes a certain time. This means that definition of such net would take into account the timed characteristics described on the transitions

$$S_T = \{P, T, I, O, H, M_0, \tau\} \tag{6}$$

where $S_M = \{P, T, I, O, H, M_0\}$ marked Petri net, $\tau : T \rightarrow \mathbf{R}_+$ – delay function, specifying static delay $\tau(t)$ of transition t.

Characteristics on transitions may determine time associated with firing of the transition in different ways. In particular, this value may be described by a deterministic or a random variable with a given probability distribution. In the latter case, we may talk about the stochastic network. In addition to static delay it is sometimes convenient to use dynamic delay $\delta(t)$, defined as the rest of the time remaining until the firing of the transition t.

In timed Petri nets, the problem of verifying the conditions required for activation of transitions is closely related to treatment of transitions that have not been fired due to the expiration of the time less than $\tau(t)$, and which had lost activity. Depending on the specific system being modelled, there are three approaches possible:
- lack of memory – after firing of any transition, dynamic delays for all transitions are set back to the initial value, i.e. $\forall t \in T, \delta(t) = \tau(t)$,
- active memory – in case of firing any transition t, all other transitions, which lost activity as a result, shall take the value of dynamic delay equal to the initial value (as in the lack of memory case), and the transitions that remain active - will retain their existing value of $\delta(t)$,
- absolute memory – no matter which transition fires, all other transitions retain their dynamic delay value, and at next activation, countdown of the time remaining for firing continues.

2.6 Coloured Petri net

The main difference between generalised and coloured nets is the ability to define markers of different types. This is possible in coloured Petri nets (CPN). Marker type is called a colour. Each place in the coloured net is assigned a set of colours that it can store. Expressions are assigned to arcs and transitions that allow manipulating various types of markers. Coloured Petri net can be written as

$$S_C = \{\Gamma, P, T, I, O, H, C, G, E, M_0\} \qquad (7)$$

where $S_M = \{P, T, I, O, H, M_0\}$ marked Petri net,
Γ – nonempty, finite set of colours,
C – function determining what colour markers can be stored in a given place: $C : P \to \Gamma$,
G - function defining the conditions that must be satisfied for the transition, before it can be fired; these are the expressions containing variables belonging to Γ, for which the evaluation can be made, giving as a result a Boolean value,
E – function describing the so-called weight of arcs, i.e. expressions containing variables of types belonging to Γ, for which the evaluation can be made, giving as a result a multiset over the type of colour assigned to a place that is at the beginning or the end of the arc.

2.7 Coloured, timed Petri net

It is possible to combine the idea of CPN and TPN. In this case the following structure of coloured, timed Petri net (CTPN) is formed

$$S_{CT} = \{\Gamma, P, T, I, O, H, C, G, E, M_0, R, r_0\} \qquad (8)$$

where: $S_M = \{P, T, I, O, H, M_0\}$ marked Petri net,

Γ – nonempty, finite set of colours, each of which can be timed, that means whose elements are pairs consisting of colour and a timestamp,
C, G, E – have the same meaning as in the case of CPN, but taking into account the fact that certain sets of colours can be timed,
R - set of timestamps (also called time points), closed under the operation of addition, $R \subseteq \mathbf{R}$,
r_0 – initial time, $r \in R$.

In the TPN it is necessary to implement a model clock, which defines the local time flow. This is achieved usually by using timestamps, which are generally associated with the markers. This clock is used to determine which of the transitions can be activated. The condition for activation is the existence, for all input places of the transition, markings, in which all timestamps are smaller than local time.

The timed coloured Petri net changes the meaning of the marking M, in relation to the timed colours. In this case, the marking consists of a number of markers together with their timestamps, which may be different for each of the markers.

State of the system modelled by coloured, timed Petri net is called the pair (M,r), where M is the marking and $r \in R$ is a timestamp.

2.8 Petri nets properties

For each Petri net we can determine among others: the reachability graph, reachability set, evaluate the reversibility, the presence of deadlock, liveness, and boundedness. In the presented method of analysis, the most important property of the net (modelling a traffic incident) is the reachability of selected states (markings) from initial marking M_0. It allows assessing the probability and time of transition to those selected markings. Particularly important are the dead markings, because they illustrate the situations in which we can assess whether the traffic process results in an incident or in an accident.

In many cases, the reachability graph is very complex and difficult to study, especially with the analytical methods. In those cases, methods to reduce the graph will be extremely useful (Sistla A.P & Godefroid P., 2004). The transport applications will use mostly the reduction related to stable sets of transitions. Reduction using symmetry will be used much less frequently.

3 MODELLING OF TRAFFIC INCIDENTS WITH THE USE OF PETRI NETS

As it is widely known, the traffic incidents in transport systems are almost always a result of a combination of many different factors. During the development of a dangerous situation in time, there are also inhibitory factors that hinder or prevent this process.

Transport system includes:
- passive components, namely infrastructure, including its characteristics,
- active elements, namely transport vehicles, performing tasks and creating a traffic flow,
- organisation, i.e. the relations between the elements of the transport system, aimed at realisation of transport tasks.

In this paper active elements of the transport system are studied, dealt dynamically, during the realisation of their task - that is, the traffic processes. Infrastructure and organisation are limitations to this process and must be, to some extent considered during its modelling.

The traffic process is ordered and designed to reach a specific destination of vehicles using the road (suitably organised in various branches of transport), including the organisational rules, regulations and standards to ensure the safety of all traffic participants. In this process, there are time periods in which vehicles move in a planned manner, in accordance with standard procedures. These fragments of the traffic process are characterized by its duration. The process is dynamic, because there is a change of position of vehicles in time, but from the point of view of the purpose of analysis, which is posed in this paper, it can be regarded as static. It is possible because in those time periods there are no events influencing the level of safety, and procedures such as changing speed or direction are planned, in accordance with the constraints resulting from characteristics of infrastructure components and tailored to the exploitation characteristics of vehicles.

Between these fragments there are traffic events which are extracted whereas the scope of the analysis. In the case of an analysis designed to assess the safety of the traffic process, these events are defined as having an impact on safety of traffic. For such events, one can include:
- occupation of conflicting point of the road (streets junction, runway, waterways crossing) characterised by the fact that there may be only one vehicle on it, or they may be few, but it is necessary to specify the order of passing this point by vehicles, as movement continued by each of them independently can lead to collisions,
- decision by the vehicle operator to continue the movement, or to change its parameters (direction, speed), in particular the decision to stop, or to realise an emergency manoeuvre to avoid collision,
- decision by the traffic dispatcher (air traffic controller, the railway station dispatcher, coordinator of traffic in seaport) of a similar nature,
- decision by the vehicle operator to take action that is inconsistent with the decisions (recommendations) of traffic dispatcher,
- occurrence of dynamic and intensive meteorological phenomena (storm, heavy fog), or other phenomena of an environmental nature that may affect the traffic process,
- occurrence of events (failures) associated with the vehicle or traffic control system, which cause hazard to vehicles.

The above mentioned events may have the nature of conditions, which logical value can be evaluated. In this case they are represented by a Boolean *true* or *false*. They may also have a nature of a certain process, mostly short-term. In this case, the event will be represented by its type, but also by duration.

Such an approach to the traffic process allows the use of Petri nets for modelling it. Stable traffic situations correspond to places in the net, traffic events – to transitions. Markers in places can be identified as traffic participants or states of environment. Participants may have different traffic characteristics. For example, we may consider several types of vehicles of varying size and performance. We may also consider objects constituting the disturbances, affecting the traffic process, such as pedestrians on the road, ground service cars on taxiways at the airport. Similar interpretation can be applied to states of environment or external events. Typically, these are logical conditions, and therefore existence of a marker in corresponding place represents the occurrence of the event.

As one can see the markers are of different types, which suggest the need to use coloured Petri nets. This is obviously a universal solution, but in simpler cases, the model of traffic incident can use a simpler place-transition net. This is possible if parts of the net using different types of markers are mostly disjoint. In cases where the same places are used by different types of markers CPN must be used.

Unlike other typical applications of Petri nets, in modelling traffic processes in transport, in most cases, it is necessary to use the timed Petri net. This results from the fact that time and the associated dynamic phenomena are often crucial in the analysis in this area. For example, while modelling traffic incidents, it is usually necessary to examine the time sequence of individual traffic situations, resulting in specific sequence of occupation of conflicting points. This sequence may decide about the occurrence of the accident or its avoidance. In specific cases, sometimes it is preferable to use timed characteristics associated with transitions, and sometimes associated with markers.

There is also a class of applications of Petri nets for modelling the traffic processes in transport, where it is sufficient to use non-timed nets. This is possible when considering only the sequence of events leading up to the situation of interest, or sequence of events as a consequence of certain initiating event. This is in fact a study of an event tree analysis, fault tree analysis, or bow-tie analysis. Analytical techniques derived from the theory of Petri nets, applied in this case, can produce very interesting results; in particular, can accelerate obtaining satisfactory results with high accuracy.

This paper describes two examples traffic incidents examination. First one is air traffic incident, with particular emphasis on modelling the process of transformation from the incident to an accident. Stochastic TPN with time associated with transitions was used. The second one is a model of waterway intersection, where two conflicting traffic flows occur. In this case stochastic CTPN was used. Term "stochastic" means that the time delays occurring in the net are partially random values of given probability distributions. Network structure itself, however, is deterministic.

4 EXAMPLE ANALYSIS – SERIOUS AIR TRAFFIC INCIDENT 344/07

As an example illustrating the method a serious air traffic incident, which occurred in August 2007 at Warsaw airport will be presented. Its participants were Boeing 767 and Boeing 737 aircraft, and its cause was classified as a "human factor" and the causal group H4 - "procedural errors" (Civil Aviation Authority 2009).

4.1 Circumstances of the serious incident

In the incident on 13th of August 2007 participated two aircraft – Boeing 737 (B737) and the Boeing 767 (B767), which more or less at the same time were scheduled for take-off from the Warsaw-Okęcie airport. As the first, clearance for line-up and wait on runway RWY 29 was issued to B737. As a second, clearance for line-up and wait on runway RWY 33 was given to B767 crew. The latter aircraft was the first to obtain permission to take-off. A moment after confirmation of permission to take-off, both aircraft began start procedure at the same time. B737 crew assumed that the start permission was addressed to them. They probably thought that since they first received permission to line up the runway, they are also the first to be permitted to start. An air traffic controller (ATC) did not watch planes take-off, because at this time he was busy agreeing helicopter take-off. The situation of simultaneous start was observed by the pilot of ATR 72, which was standing in queue for departure. He reacted on the radio. After this message, B767 pilot looked right and saw B737 taking-off. Then, on his own initiative, broke off and began a rapid deceleration, which led to stopping the plane 200 meters from the intersection of the runways. Assistant controller heard the ATR 72 pilot radio message and informed the controller that B737 operate without authorization. A controller, who originally did not hear the information by radio, after 16 seconds from the start, recognized the situation and strongly ordered B737 to

discontinue take-off procedure. B737 crew performed braking and stopped 200 m from the intersection of the runways.

4.2 Model of serious incident

This air traffic incident almost led to collision between the two aircraft, it means to accident. As in most such situations, there were many factors contributing to the creation of this dangerous situation. The most important are:
- lack of situational awareness at the B737 crew,
- inadequate monitoring of radio communications and, consequently, wrong acceptance of permission for the start, in fact directed to another plane,
- lack of the crew cooperation in the B737 cockpit,
- lack of proper monitoring of the take-off by the controller,
- controller's lack of response to the information from the pilot of ATR 72 transmitted by radio.

The factors impeding the development of the accident, which resulted in preventing it, include:
- good assessment of dangerous situation by the crew of B767 and decision to immediately discontinue take-off,
- good recognition of the hazard by the crew of the ATR 72 and immediate sending a message by radio,
- good weather conditions for visual observation of the runways,
- proper response of assistant controller.

TPN model representing this serious incident is shown in Figure 1.

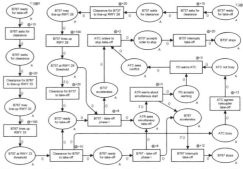

Figure 1. The basic model of a serious air traffic incident 344/07

4.3 Model of air traffic accident

Analysis of the factors leading to the incident may give an answer to the question what is the probability of such incident. For example, one may check how the situation would change if it was B767 the first aircraft to obtain permission to line up the runway.

In the presented example, however, a goal is to find a probabilistic dependence between the serious incident and an accident that could result from it. In this case, it is necessary to notice that it is sufficient that there exists only one additional factor, and incident would in fact be an accident. There are several scenarios that lead to an accident.

1 B767 crew, busy with their own take-off procedure does not pay attention to the message transmitted by radio by the ATR 72 pilot.
2 B767 crew takes a wrong decision to continue the take-off, despite noting B737 aircraft.
3 ATR 72 pilot does not watch the situation on the runways, just waiting for permission to line-up the runway.
4 ATR 72 pilot observes a dangerous situation, but does not immediately inform about it on the radio, instead discusses it with other members of his own crew.
5 Assistant controller does not pay attention to the information given by radio by the ATR 72 pilot, or does not respond to it properly - does not inform the controller.
6 Weather conditions (visibility) are so bad that it is impossible to see the actual traffic situation. This applies to B767, ATR 72 crews, and the air traffic controller.

All these scenarios will lead with certainty (or with great probability) to transformation of the incident into an accident, and can be analyzed using Petri net model. In this analysis one should take into account the possibility of occurrence of each scenario separately, as well as several of them at once.

4.4 Probability of incident-accident transformation

Analysis of the probability of transformation of incident into an accident must take into account the probability of each scenario mentioned above. In the case of scenario 6 we can use statistical data on meteorological conditions (visibility) in the airport. But in other scenarios, it is necessary to refer to experts' evaluation.

Taking into account the objectives of the analysis, it is possible to eliminate certain states without loss of accuracy, while simplifying the analyzed model. This applies, for example, to almost all the places and transitions associated with the process of taxiing and lining up the runway. For example, change the set of places is determined as follows.

$$P_w = (P - P_r) \cup P_d \qquad (9)$$

where: P_w - a set of places in the modelled accident,
P_r - a set of reduced places,
P_d - a set of places added to the model, to reflect the above-mentioned scenarios.

In this case (Figure 1) P_r = {"B767 awaiting permission to start", "B767 can line up RWY 33", "B767 on the RWY 33 threshold", "B767 ready for

take-off", "B737 awaiting permission to start", "B737 can line up RWY 29", "B737 on the RWY 29 threshold", "B737 ready for take-off", "ATC not busy", "ATC busy", "ATR observes a simultaneous start"}.

On the other hand P_d = {"ATR warns?", "B737 continues to start", "B737 at the crossing", "B767 hears the warning?", "B767 continues to start", "B767 at the crossing", "B767 interrupts start?", "B767 begins deceleration", "weather?", "good visibility"}.

A similar modification was made in regard to transitions, input, output and inhibition functions. An additional issue to consider is change of transition type – from timed to immediate or vice versa. Petri net to model the transformation of the incident into accident, after reduction is shown in Figure 2.

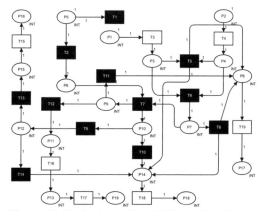

Figure 2. Model of serious incident 344/07 transformation into air traffic accident (after reduction of the states).

This network may be treated as a stochastic timed Petri net. Its analysis allows observing some interesting relationships between a serious incident and the air traffic accident. It also allows determining some quantitative dependencies.

Assume the following places designations: p_1 – „B767 ready for take-off", p_2 – „B737 ready for take-off", p_3 – „B767 accelerates", p_4 – „B737 accelerates", p_5 – „weather?", p_6 – „good visibility", p_7 – „ATR warns?", p_8 – „B737 continues take-off", p_9 – „FD accepts warning?", p_{10} – „B767 hears warning?", p_{11} – „ATC sees conflict", p_{12} – „B767 interrupts take-off?", p_{13} – „B737 accepts order to interrupt take-off", p_{14} – „B767 continues take-off", p_{15} – „B767 begins braking", p_{16} – „B767 stops", p_{17} – „B737 at crossing", p_{18} – „B767 at crossing", p_{19} – „B737 stops".

The set of all states, called a reachability set, for model of accident is presented in Table 1.

The most important markings, from the perspective of the analysis presented in this article, are given in Table 2. Other states as well irrelevant places – were omitted.

Table 1. The reachability set for the model of accident arising from incident 344/07

M_0	$p_1+p_2+p_5$	M_1	p_1+p_2	M_2	$p_1+p_2+p_6$
M_3	p_2+p_3	M_4	p_1+p_4	M_5	$p_2+p_3+p_6$
M_6	p_4+p_6	M_7	p_3+p_4	M_8	$p_3+p_4+p_6$
M_9	p_8+p_{14}	M_{10}	p_7	M_{11}	$p_6+p_8+p_{14}$
M_{12}	p_6+p_7	M_{13}	p_8+p_{18}	M_{14}	$p_{14}+p_{17}$
M_{15}	p_8+p_{18}	M_{16}	$p_6+p_{14}+p_{17}$	M_{17}	p_9+p_{10}
M_{18}	$p_{17}+p_{18}$	M_{19}	$p_6+p_{17}+p_{18}$	M_{20}	$p_{10}+p_{11}$
M_{21}	p_8+p_{10}	M_{22}	p_9+p_{12}	M_{23}	p_9+p_{14}
M_{24}	$p_{11}+p_{12}$	M_{25}	$p_{11}+p_{14}$	M_{26}	p_8+p_{12}
M_{27}	p_9+p_{15}	M_{28}	$p_{11}+p_{15}$	M_{29}	$p_{11}+p_{18}$
M_{30}	$p_{13}+p_{14}$	M_{31}	p_8+p_{15}	M_{32}	$p_{11}+p_{16}$
M_{33}	$p_{13}+p_{15}$	M_{34}	$p_{13}+p_{18}$	M_{35}	$p_{14}+p_{19}$
M_{36}	p_8+p_{16}	M_{37}	$p_{15}+p_{17}$	M_{38}	$p_{13}+p_{16}$
M_{39}	$p_{15}+p_{19}$	M_{40}	$p_{18}+p_{19}$	M_{41}	$p_{16}+p_{17}$
M_{42}	$p_{16}+p_{19}$				

Table 2. Selected states of the system (model of accident)

	M_{18}	M_{19}	M_{40}	M_{41}	M_{42}
p_6 – good visibility	0	1	0	0	0
p_{16} - B767 stops	0	0	0	1	1
p_{17} - B737 at crossing	1	1	0	1	0
p_{18} - B767 at crossing	1	1	1	0	0
p_{19} - B737 stops	0	0	1	0	1

States M_{40}, M_{41}, M_{42} (called safe states) illustrate situations in which there is no accident. States M_{18} and M_{19} represent the situation that analysed serious incident transforms into accident. The joint probability of finding system in one of these states is the searched probability of incident-accident transformation. It can be determined both analytically and by simulation using a suitable software tool. Analytical method for determining the sought probabilities will be presented on the example of the final state M_{19}. Partial subgraph of the reachability graph, for reaching M_{19} from initial state M_0 is shown in Figure 3.

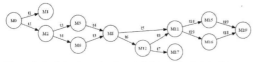

Figure 3. Partial subgraph of reachability of final state M_{19}.

Let's assume the following transitions designations: t_1 – „bad weather", t_2 – „good weather", t_3 – „B767 take-off phase I", t_4 – „B737 take-off phase I", t_5 – „ATR not watches", t_6 – „ATR watches", t_7 – „ATR warns", t_8 – „ATR not warns", t_9 – „B767 hears", t_{10} – „B767 not hears", t_{11} – „FD not accepts", t_{12} – „FD accepts", t_{13} – „B767 interrupts", t_{14} – „B767 not interrupts", t_{15} – „B767 decelerates", t_{16} – „ATC orders B737 to interrupt", t_{17} – „B737 interrupts take-off and stops", t_{18} – „B767 take-off phase II", t_{19} – „B737 take-off phase II".

Immediate transitions t_1, t_2, t_5, t_6, t_7, t_8, t_9, t_{10}, t_{11}, t_{12}, t_{13}, t_{14} are assigned weights, respectively: α_1, α_2, α_5, α_6, α_7, α_8, α_9, α_{10}, α_{11}, α_{12}, α_{13}, α_{14}. These weights are used to determine the probability of firing transitions in a situation of a conflict. Timed transitions t_3,

t_4, t_{15}, t_{16}, t_{17}, t_{18}, t_{19} are assigned the intensities of realisation, respectively: μ_3, μ_4, μ_{15}, μ_{16}, μ_{17}, μ_{18}, μ_{19}. Also for this type of transitions in the event of a conflict, it is necessary to determine the probability of firing one of the conflicting transitions.

Because of the purpose of analysis, it is possible to reduce the reachability graph. Reduction consists of the removal of states that do not affect the probability of finding the system in the state M_{19}. Reachability graph after reduction is shown in Figure 4.

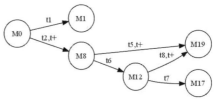

Figure 4. Reduced reachability graph for state M_{19}.

In this case, the probability that the system will move from the state M_0 to M_{19} depends on the probabilities of firing of immediate transitions t_2, t_5, t_6 and t_8, and is described by the two sequences σ_1 and σ_2, and after reduction of intermediate states is as follows:

$$\sigma_1 = M_0\left[t_2,t_+\right\rangle M_8\left[t_5,t_+\right\rangle M_{19} \qquad (10)$$

$$\sigma_1 = M_0\left[t_2,t_+\right\rangle M_8\left[t_6\right\rangle M_{12}\left[t_8,t_+\right\rangle M_{19} \qquad (11)$$

$$P\left(M_0\left[\sigma_{1-2}\right\rangle M_{19}\right) = \frac{\alpha_2}{\alpha_1+\alpha_2} \cdot \left(\frac{\alpha_5}{\alpha_5+\alpha_6} + \frac{\alpha_6}{\alpha_5+\alpha_6} \cdot \frac{\alpha_8}{\alpha_7+\alpha_8}\right) \qquad (12)$$

It is worth noting that in this case the probability of transforming incident into accident is not affected by intensities of timed transitions, and only the weights of immediate transitions.

5 EXAMPLE ANALYSIS – VESSEL TRAFFIC AT WATERWAYS INTERSECTION

Majzner & Piszczek (2010) formulated the interesting problem of analysis of traffic safety at the intersection of the waterways. This problem can be modelled using the presented method.

Two streams of traffic are studied: longitudinal moving along the fairway with the speed v_w and crossing stream moving with the speed v_p. It was assumed that the ships moving in the longitudinal stream have the right of way to the ships in the crossing stream. The study analyses the average waiting time for ships of crossing stream and the probability of avoiding a premise for a collision, as a function of intensity of longitudinal stream. Example of Petri net for modelling this kind of problem is presented in Figure 5. The net is the coloured, sto-chastic, timed Petri net with priorities.

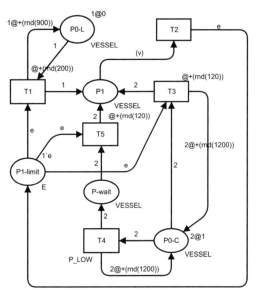

Figure 5. CPN modelling incidents at waterway intersection.

Places designations are: p_{0L} – "unit from longitudinal stream arrives at intersection", p_{0C} – "unit from crossing stream arrives at intersection", p_1 – "vessel occupies intersection", $p_{1\text{-limit}}$ – "anti-place for limiting the number of vessels at intersection to 1", p_{wait} – "vessel waits in the queue".

Transitions designations are: t_1 – "unit from longitudinal stream enters the intersection", t_2 – "unit leaves the intersection", t_3 – "unit from crossing stream enters the intersection", t_4 – "unit from crossing stream enters the waiting area", t_5 – "unit from waiting area enters the intersection",

Assuming the figures from discussed example we obtain similar results. For example, for the parameters shown in Figure 5 (longitudinal traffic - 4 units per hour, crossing traffic - 3 units per hour) consistency of results from simulation experiments with the results of the sample model is above 90% for mean delay time.

This indicates the usefulness of the proposed modelling method to analyze safety and traffic capacity problems in the fairways. We may also expect good results while researching other problems in the field of maritime traffic engineering.

6 SUMMARY AND CONCLUSIONS

In the paper the method of modelling traffic incidents and accidents was presented.

Petri nets are used for modelling. Type of net used, depends on individual case and objective of analysis. Presented examples show the applicability of the proposed method for analysis of traffic processes in various modes of transport. The use of Petri nets allows to easily generating the reachability graph, which is the basic tool of analysis. This graph is often large and the effective application of the method depends on its reduction. The problem of effective reduction constitutes a different research problem.

Method can be used in practice for improvement of the transport safety. The case described as aviation example is a part of analysis that is necessary before any new equipment or procedure can be introduced. The simple model described as maritime example may be used as a part of more complex optimisation models for marine traffic engineering.

REFERENCES

Aviation Law 2002. Act of 3 July 2002 (Journal of Laws of 2002, No. 130, item. 1112) (in Polish)

Civil Aviation Authority 2009. Statement No. 78 of President of the Office of Civil Aviation from 18 September 2009 on air event No. 344/07, Warsaw. (in Polish)

Jensen K. 1997. Coloured Petri Nets. Basic Concepts, Analysis Methods and Practical Use. *Monographs in Theoretical Computer Science.* Springer Verlag.

Majzner P. & Piszczek W. 2010. Investigation of vessel traffic processes at waterway intersections, *Scientific Journals Maritime University of Szczecin,* vol. 21(93)/2010, p. 62-66. Szczecin.

Marsan M.A., Balbo G., Conte G., Donatelli S., Franceschinis G. 1999. *Modelling with Generalized Stochastic Petri Nets,* Universita degli Studi di Torino, Dipartimento d'Informatica.

Sistla A.P. & Godefroid P. 2004. Symmetry and reduced symmetry in model checking, *ACM Transactions in Programming Languages Systems,* 26(4) p. 702-734.

Skorupski J. 2010. Simulation analysis of relation between serious incident and accident in air traffic, *Logistics* (ISSN 1231-5478) No. 4/2010, Institute of Logistics and Warehousing, Poznan.

Szpyrka M. 2008. Petri Nets in Modelling and Analysing of Concurrent Systems, WNT, Warszawa. (in Polish)

4. Fire Safety Assessment of Some Oxidizers in Sea Transport

K. Kwiatkowska-Sienkiewicz
Gdynia Maritime University, Poland

ABSTRACT: Oxidizers belong to 5.1 Class danger goods according to International Maritime Dangerous Goods Code. This paper provides an outlook on fire safety assessment concerning some oxidizers in sea transport. The investigation was aimed at comparison of two methods of classification of these materials. First research was conducted in accordance with the test described in IMDG Code. The second method was the differential thermal analysis (DTA) where the basis was the determination of the temperature change rate during thermal reactions. On this base investigated substances were assigned to a packing group of dangerous goods. According to two used tests, the investigated oxidizers belong to 5.1 class, including magnesium nitrate, and require to packaging group III or II packaging group of the International Maritime Danger Goods Code. The DTA method gives more quantitative information about fire risk on the ship than method recommended in International Maritime Danger Goods Code.

1 INTRODUCTION

The new trend in international regulations emphasizes the importance of qualitative risk assessment. The sea transportation of dangerous materials is connected which certain risk. During last decade attention of marine world has been focused on safety of shipping. The number of accidents has increased almost twice from the year 2000. The reason of this effect is not clear – it may be connected with increase of the size of ships and other factors, but most probably to the lower level of performance of crew members which were recruited from many different countries. A great number of accidents on the sea are caused by wrong management, bad organization and crew error [Kobyliński, 2009].

The Chemical Abstract Service (CAS) lists over 63 00 chemicals used outside the laboratory environment, and number increases each year. In the mid - 1970s the United States Department of Transportation (DOT) established a definition of hazardous material. DOT began the first major regulation of hazardous materials in transportation, including a hazard class and placard and label system for identification of those materials. In that time began developing regulations dealing with hazard materials behavior, storage, and use [Burke, 2004].

Nowadays we need to observe more consideration of aspects of transportation of large quantity of danger-

ous goods. The term dangerous is limited to substances which have the potential to cause major accident risk from fire, explosion or toxic release.

In the International Maritime Dangerous Goods Code the information concerning various aspects of sea handling of dangerous goods is contained. Official regulations and supplementary documentation of the hazardous properties of materials can be found in this code. It is an important source of basic informed and a guide to shipping of dangerous goods for a ship staff.

Oxidizers are dangerous goods in accordance with International Maritime Dangerous Goods Code (IMDG-Code), belong to 5.1 class dangerous goods. They are not necessarily flammable, but able to intensify the fire by emission of oxygen [IMDG – Code, 2004]. Oxidizers may be elements, acids, salts and organic compounds, generally - liquids or solid substances. Oxidizing materials, during transport might be initiated by fire, impact. Some oxidizing substances have toxic or corrosive properties, or have been identified as pollutant to the marine environment. They will react in contact with reducing reagents. Hence oxidizing agent will invariably accelerate the rate of burning of combustible material. The National Fire Protection Association in the United Stated classified oxidizing substances according to the stability in four class [Burke, 2004]:

1 solid or liquid that readily yields oxygen or oxidizing gas or that readily reacts to oxidizer combustible materials,

2 oxidizing material can cause spontaneous ignition when contact with combustible materials,
3 oxidizing substances that can undergo vigorous self sustained decomposition when catalyzed or exposed to heat,
4 oxidizing articles that can undergo an explosive reaction when catalyzed or exposed to heat, shock or friction.

A major fire aboard a ship carrying oxidizing materials may involve a risk of explosion in the event of contamination by combustible materials. An adjacent detonation may also involve a risk of explosion. During thermal decompose oxidizers giving toxic gases and gases which support to combustion. Dust and gasses might be irritating to skin and mucous membranes.

Before loading of oxidizers, attention must be paid to the proper cleaning of holds or compartments into which they will be loaded. In the event of fire involving substances of this Class oxygen is liberated which has the result of making the fire self sustaining even in completely inert atmosphere.

Many dangerous goods should be effectively segregated from oxidizers. There are four steps of segregation of that incompatibles substance [IMDG – Code, 2004]:

Away from - inflammable solids (Class 4.1), poisons (Class 6.1), radioactive substances (Class 7);

Separated from – explosives (Class 1, Divisions 1.4 and 1.5), inflammable gases (Class 2), inflammable liquids (Class 3), spontaneously combustible substances (Class 4.2), substances which are dangerous when wet (Class 4.3), organic peroxides (Class 5.2), corrosives (Class 8);

Separated by a complete compartment or hold from cargos of fibrous materials;

Separated longitudinally by an intervening complete compartment or hold – explosives (Class 1, Divisions 1.1, 1.2, 1.3).

If oxidants are very dangerous, may be transported in limited quantities. Some of substances packaging group II may have maximum net quantity per inner receptacle - 50g, maximum net quantity per package – 1kg and maximum net quantity which may be shipped from one consignor to one consignee -10 kg.

For materials packaging group III – maximum net quantity per inner receptacles are 500g, maximum net quantity per package – 10kg and maximum net quantity which may be shipped from one consignor to one consignee -50 kg. The competent authority may allow exemptions from the provisions.

Fire protection of oxidizers is the general problem with those cargos.

There are at present not established good criteria for determining packaging groups and segregation of oxidizers. To packaging group I belongs substances great danger, II - medium danger, or III, minor danger. The assignation criteria to the packaging groups and segregation from incompatible substances are based on a physical or chemical property of goods.

Classification of oxidizing substances to class 5.1 is based on test described in the IMDG Code and Manual of Tests and Criteria [UN Recommendations Part III]. In this test, the investigated substances were mixed with cellulose. The mixtures were ignited and the burning time trials forms were noted and compared to burning time a reference mixture - potassium bromate(V) and cellulose.

Cellulose belongs to polysaccharides; develop free radicals on heating [Ciesielski, Tomasik 1998, Ciesielski, Tomasik, Baczkowicz 1998]. Free radical exposed during thermal reaction polysaccharides mixed with oxidizers gives possibility of self-heating and self-ignition chemical reaction.

In practice, during long transport combustible materials and commodities containing polysaccharides we can observe self-heating effect, specially, if polysaccharides are blended with oxidizers or explosives substances.

Using differential thermal analysis (DTA) we can registr quality and quantity changes during dynamic heating of investigated materials in time.

The self-heating or thermally explosive behavior of individual chemicals is closely related to the appearance of thermogravimetry-differential thermal analysis (TG-DTA) curve with its course.

In previous examinations of mixtures of oxidizers with cellulose and flour wood the temperature change rates [^{0}C/s] were calculated into 1 milimole of an oxidizer and tested oxidizing substances were blended with combustible substance in mass ratio 5:1 [Michałowski, Rutkowska, Barcewicz 2000, Kwiatkowska-Sienkiewicz et al. 2006, Kwiatkowska-Sienkiewicz 2008].

Determination according to IMDG Code and UN Recommendations test and DTA method were carried out using the same blends oxidants and cellulose [Kwiatkowska-Sienkiewicz, Kałucka 2009, Kwiatkowska-Sienkiewicz 2010].

In this paper we concerned comparison of two methods of assignation to class 5.1 and classification to packing groups and separation from incompatible substances. The first one is recommended by the United Nations [UN Recommendations Part III] and second one, differential thermal analysis is used in chemistry. Recommended UN standards for fire protection of oxidizers give only qualitative information about time of fire transfer along the oxidizer-cellulose mixture. DTA gives quantitative information about time and temperature changes during all processes of selfheating and selfignition. We can find warmth quantitative determination.

2 EXPERIMENTS

2.1 *Examination of burning time*

Classification of oxidizing goods to class 5.1 of IMDG – Code is based on test described in the Manual of Tests and Criteria [UN Recommendations Part III].

In this test, the examinee substances were mixed with cellulose in ratios of 1:1 and 4:1, by mass, of substance to cellulose. The mixtures were ignited in standard form and the burning time was noted and compared to burning time a reference mixture of potassium bromate(V) and cellulose, in ratio 3:7, by mass.

Any substance which, in both the 4:1 and 1:1 sample-to-cellulose ratio (by mass) tested, does not ignite and burn, or exhibits mean burning times greater than that of a 3:7 mixture (by mass) of potassium bromate(V) and cellulose, is not classified as class 5.1.

If a mixture of test substance and cellulose burns equal to or less than the reference mixture (bromate(V) : cellulose = 3:7) this indicates that substance has oxidizing (fire enhancing) properties and is classified in class 5.1.

This also means that oxidizing substance is assigned to a packing group III (if the criteria of packing group I and II are not met).

Next the burning time is compared with those from the packing group I or II reference standards, 3:2 and 2:3 ratios, by mass, of potassium bromate(V) and cellulose.

Mean burn time of trials mixtures oxidizers and cellulose are presented in Table 1.

2.2 *Differential thermal analysis*

Thermal reaction course during the heating can be investigated by means of differential thermal analysis (DTA) method. Using thermal analysis (DTA), the changes of mass, temperature and heating effects curves are recorded.

The following outputs were recorded during measurements using DTA method:
– the temperature change curve which is a straight line till the mixture flash point is reached, with a district peak in the self-ignition region, especially during reaction of very active oxidizers,
– DTA curve – gives information about heat effects,
– curve of mass change during the reaction.

The temperature increase value was determined on the basis of its deflection out of the straight line, in the peak region. On the ground of the above mentioned data following parameters could be calculated: the temperature change rates [$^{\circ}$C/s], were calculated by dividing the temperature increase by the

time of self-ignition effect, counted into 1mol of oxidizers.

The examination basing on potassium bromate(V) blends with cellulose (in mass ratio 2:3 and 3:7) as a standard shows that class 5.1 includes substances which temperature change rate was equal to or less than the reference temperature change rate of mixture of potassium bromate (V) with cellulose, in mass ratio 3:7 – 0,96 [$^{\circ}$C/s].

To the III packaging group should be assigned substances, which during thermal analysis mixtures oxidants with cellulose, the temperature change rate values are between 0,96– 1,82 [$^{\circ}$C/s].

The II packaging group involves crossing value of the temperature change rate under 1,82 [$^{\circ}$C/s].

2.3 *Apparatus and reagents*

About 30 g samples were used to prepare trial form to examination burning time, according to IMDG Code.

Thermal treatment of pure oxidizers or the blends with cellulose were heated from room temperature to 500°C or 1000°C. The procedure was run in the air, under dynamic condition. The rate temperature increase was 10°C/min. Ceramic crucibles were taken. Paulik-Paulik-Erdley 1500 Q Derivatograph (Hungary) was used. Samples used in DTA method were 300 - 600 mg. The measurements were carried out three times. Decomposition initiating temperatures of the compounds and his blends with cellulose were read from the recorded curves.

In experiments used microcrystalline cellulose, grade Vivapur type 101, particle size >250 µm (60 mesh), bulk density 0.26 – 0.31 g/ml.

The following substances were blended with cellulose in mass ratio 1:1 or 1:4
– lead nitrate(V), pure for analysis,
– cadmium nitrate{V), pure for analysis,
– silver nitrate(V),), pure for analysis,
– sodium dichromate, pure for analysis,
– potassium dichromate, pure for analysis.
As standard used potassium bromate(V), pure for analysis.

3 RESULTS

The results of performed thermal reactions of oxidizers and cellulose are presented in Tables 1, 2 and 3.

Mean data of trials mixtures oxidizers and cellulose are presented in Table 1.

On the basis of results of the test described in Manual of Test and Criteria examinee fire oxidizers, according to IMDG Code, belong to class 5.1 of dangerous goods, but according to investigation, magnesium nitrate(V) does not belong to oxidizers. Sodium and potassium dichromates according to

IMDG Code have not oxidizing properties, after experiments they are found in 5.1 Class and require packaging group III. In case lead nitrate(V), the investigation are agreeable with IMDG Code. For silver nitrate(V), the reduce packaging group from II to III has been proposed.

Using these criteria big mass samples of components involve larger volumes of toxic gases during heating.

The results of second method of performed thermal reactions between cellulose and selected oxidizers are presented in Tables 2 and 3.

During thermal reactions there were observed exothermic processes self-heating, self-ignition and weight losses. A weight loss is very dangerous in shipping, especially of bulk cargo. Oxidizers containing nitrates and dichromates during fire on the boat, lost stowage mass about 1/3.

The results of differential thermal analysis suggest similar effects like in the tests recommended by IMDG Code; all investigated oxidizers belong to 5.1. class of dangerous goods and require packaging group III including lead and cadmium nitrates(V), where gives self-ignition effects and proposed for them packaging group II.

Table 1. Determination risk of fire oxidizers according to Manual Test and Criteria IMDG Code

Oxidizer	Sample to cellulose.	Burn rate [cm/min.] Samppl e	Stand ard	Proposed Class IMDG	Pacaging-group
Lead nitrate(V)	1:1	10,4	0,83	5.1	II
	4:1	9,1			
Cadmium nitrate(V)	1:1	2,3	0,83	5.1	III
	4:1	0,94			
Silver nitrate(V)	1:1				
	4:1	1,05	0,83	5.1	III
Barium nitrate(V)	1:1	2,18			
	4:1	0	0,83	5.1	III
Sodium dichromate	1:1	2,70			
	4:1	9.1	0,83	5.1	III
Potassium dichromate	1:1	1,01	0,83	5.1	III
	4 :1	0,75			
Potassium bromate	2:3	10	10	5.1	II
(V) (p.a.) (standard)	3:7	0,83	0,83	5.1	III

Table 2. Thermal decomposition oxidizers and his blends with cellulose using DTA method

Oxidizers	Sample to cellulose	Ignition temperature [°C] Oxidizer	Ox. – cell.	Temperature change rate [°C/s] Ox.	Ox.-cell.	Thermal effects
Lead nitrate(V)	1:1		183			
					1.12	+
	4:1	425		0,5		
			211		1,32	+
Cadmium nitrate(V)	1:1		197,3			
					1.6	+
	4:1	375		0,5		
			218,3		1,74	+
Silver nitrate(V)	1:1					
	4:1	530	188	0,2	3,41	+
Barium nitrate(V)	1:1		321			+
		758		0,3	0,99	
	4:1		343			+
					0	
Sodim dichromate	1:1		133		1,0	+
		349		0,1		
	4:1		145		0,2	+
Potassium dichromate	1:1		129		1,1	+
		295		0,2		
	4:1		150		0,4	+
Potassium bromate (V) (p.a.) (Standard)	3:7				0,96	+
		455	329,3			
	2:3		190	2,6	1.82	+

Ox. – oxidizer
cell. – cellulose
+ exothermic effect of process

Blends of potassium bromate(V) and cellulose in mass ratio 3:7 and 2:3 are too the standards in classification using differential thermal analyses tests (used the same standards like in Manual Test recommended by IMDG Code).

The ignition temperature and temperature change rate make it possible to assess packaging group of investigated oxidizers, belong to class 5.1 of danger goods. The blends of oxidizers and cellulose had lower ignition temperature than pure oxidizers. Pure nitrates(V) and chromates, high ionic compounds, had thermal decompose in higher temperatures than 500^0C, but the mixtures with combustible material – cellulose were decomposed in temperature about 200^0 C.

Table 3. Assignment of the oxidizers to the packaging group based on differential thermal analyze (DTA)

Name of the oxidizer	Temperature change rate [°C/s]	Assigned packaging group	Proposed class of IMDG Code
Lead nitrate(V)	1,32	III	5.1
Cadmium nitrate(V)	1,74	I/II	5.1
Silver nitrate(V)	3,41	III	5.1
Barium nitrate (V)	2,1	III	5.1
Sodium dichromate	1,0	III	5.1
Potassium dichromate	1,1	III	5.1
Potassium bromate (V) (Standard)	0,96÷1,82	III	5.1

After comparison these two methods of assignation to class 5.1 and packaging group's data thermal analysis gives quantitative information about thermal effects during heating. This method gives possibility to denote melting, self-heating, self-ignition temperatures and loss mass during heating. During Manual Test (according of IMDG Code) we have only qualitative data burning time of blends oxidizers with cellulose. In Manual Test was used big probe - 30g of blends oxidizer and cellulose, in DTA method only - 300 ÷500 mg.

Differential thermal analysis is objective chemical method which could make it possible to determine the criteria of assignment of oxidizers to packaging groups, required for sea transport. Data DTA method gives more information about fire risk assessment that Manual Test recommended by IMDG Code. From thermal analysis we can have also information about volumes of toxic gases diffusions during fire (e.g. NO_x, CO, CO_2).

4 CONCLUSION

– Manual Test recommended in IMDG Code informs only about qualitative fire risk,
– data based on the differential thermal analysis gives more information about fire risk assessment than Manual Test recommended by IMDG Code,

– the comparison of two methods of classification and assignment to a packing group of solid substances of class 5.1 of IMDG Code indicate, that differential thermal analysis (DTA method) gives objective, quantitative information about fire risk on the boat. Using this method, during heating we can register changes of temperatures, melting point, self-heating, self-ignition explosive effects and determination of mass loss,
– differential thermal analysis method should be found in the standards of investigations of fire protection oxidizers.

REFERENCES

Bruke R.., 2003; *Hazardous Materials Chemistry for Emergency Responders.*, Bocca Ratton-London-New York-Washington. :Lewis Publishers CRC Press Company,
Ciesielski W., Tomasik P., 1998; .Starch radicals. Part III: Semiartifical complexes. Z. Lebensm. Untes. Forsch. A. 207:pp. 292-298.
Ciesielski W., Tomasik P. Baczkowicz M.,1998; Starch radicals. Part IV: Thermoanalitical studies.. Z. Lebensm. Untes. Forsch. A. 207:pp. 299-303.
International Maritime Dangerous Goods Code (IMDG Code) 2008; London, Printed by International Maritime Organization.
Kobyliński L., 2009; Risk analysis and human factor in prevention of CRG casualties, *Marine Navigation and Safety of Sea Transportation* - Weintrit ed., London, pp. 577-582
Kwiatkowska-Sienkiewicz K., Barcewicz K., Rojewski L., 2006; Application of thermal analyze in studies on sea transport safety of oxidizers (3). *Proc.The 15 th Symposium of IGWT*, Kyiv, Ukraine , pp. 1182-1186.
Kwiatkowska-Sienkiewicz K., 2008; Application of thermal analyze in studies on levels of separation of oxidizers from ammonium chlorate(VII).*Proc. The 16 th Symposium of IGWT*, Suwon, Korea, pp. 737-739.
Kwiatkowska-Sienkiewicz K., Barcewicz K.., 2001; The new criteria of separation of oxidizers from ammonium salts. *Proc. European Safety & Reliability Conference, ESREL*, Turyn, Italy, Ed. Technical University of Turyn, pp.959-963.
Kwiatkowska-Sienkiewicz K., Kałucka P., 2009; Application of thermal analysis and trough test for determination of fire safety on some fertilizers containing nitrates, *Marine Navigation and Safety of Sea Transportation* - Weintrit ed., London, pp. 651-655
Kwiatkowska-Sienkiewicz K., 2010; Principles of Fire Risk Assessment in sea Transportation of Nitrate Goods Safety, *Polish Journal of Environmental Studies*, 19 (4A), 69-72.
Manual of Tests and Criteria Part III, 34.4.1.: 2001: London, United Nations Recommendations on the Transport of Dangerous Goods.

Information and Computer Systems in Transport Process

5. Development and Standardization of Intelligent Transport Systems

G. Nowacki

Motor Transport Institute, Warsaw, Poland

ABSTRACT: The paper refers to theoretical basis and history of Intelligent Transport Systems. The first term telematics was created in 1978, then transport telematics in 1990 and term - Intelligent Transportation Systems (ITS) were approved in USA and Japan in 1991 and in Europe in 1994 on the world ITS Congress in France. The development and standardization of Intelligent Transportation Systems has been presented. ITS standardization in Europe is dealt with by the following institutions: CEN, ETSI and CENELEC. Furthermore standards of the applications of maritime intelligent transport systems have been presented including maritime Management and Information Systems, sea environment and interactive data on-line networks, ship integrated decision support systems, Advanced maritime navigation services, automatic identification, tracking and monitoring of vessels, as well as safety purposes.

1 DEVELOPMENT OF ITS

1.1 *Terminology of ITS*

The term telematics comes from the French - télématique and first appeared in the literature at the end of the seventies. In 1978 two French experts: S. Nora and A. Minc, introduced this term- télématique, which was created by linking telecommunication (télécommunications) and informatics (informatique), and using the following segments of those words: télé and matique. In 1980 this term began to function also in the English terminology (Mikulski 2007). The term telematics describes the combination of the transmission of information over a telecommunication network and the computerized processing of this information (Goel 2007).

Some authors define the term telematics, as telecommunication, information and informatics technology solutions, as well as automatic control solutions, adapted to the needs of the physical systems catered for – and their tasks, infrastructure, organization maintenance processes, management and integrated with these systems (Tokuyama 1996, Piecha 2003, Wawrzyński 2003, Mikulski 2007, Nowacki 2008).

Telematics systems use various software, devices and applications:

– for electronic communication, linking individual elements of the telematics system, WAN (wide area network), LAN (local area network), mobile telecommunication network, satellite systems);
– for information collection (measurement sensors, video cameras, radars);
– of information presentation for the telematics system administrators (GIS – Geographical Information System, access control systems);
– Of information presentation for the system users (light signalling, radio broadcasting, internet technologies).

Telematics term has begun to be introduced into various branches of the economy, hence the appearance of such terms as: financial, building, health, environmental protection, operational, postal, library telematics.

A particular example illustrating the application of the telematics is modern transport. Transport telematics encompasses systems, which allow – thanks to a data transmission and its analysis – to influence the road traffic participants' behaviour or operation of the vehicles' technical elements, or out on the road, during the actual haulage (Internationales Verkehrswesen 2003).

Transport telematics term has been used in Europe since 1990.

The applications of transport telematics are Intelligent Transportation Systems (ITS).

ITS mean the systems, in which people, roads and vehicles are linked through the network utilizing, advanced information technology (Berghout & 1999).

Intelligent Transport Systems (ITS) mean systems in which information and communication technologies are applied in the field of road transport, including infrastructure, vehicles and users, and in traffic management and mobility management, as well as for interfaces with other modes of transport (Directive 2010/40/EC).

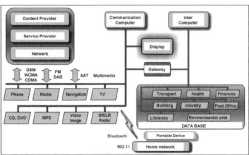

Figure.1. General structure of telematics system

Legend:
- WCDMA (Wideband Code Division Multiple Access) an ITU standard is officially known as IMT-2000 direct spread. ITU (International Telecommunication Union – former CCIT (Comité Consultatif Internationale de Télégraphie et Téléphonie) was created in the first of March 1993.
- CDMA (Code Division Multiple Access) is a spread spectrum multiple access technique. A spread spectrum technique spreads the bandwidth of the data uniformly for the same transmitted power. Spreading code is a pseudo-random code that has a narrow Ambiguity function, unlike other narrow pulse codes. In CDMA a locally generated code runs at a much higher rate than the data to be transmitted.

The general structure of Intelligent Transportation Systems applications may include: vehicle, airplane & ship operations, crash prevention and safety, electronic payment and pricing, emergency management, freeway management, incident management, information management, intermodal freight, road weather management, roadway operations and maintenance, transit management, traveller information.

Interoperability of ITS is the capacity of systems and the underlying business processes to exchange data and to share information and knowledge. ITS application means an operational instrument for the application of ITS. ITS service - the provision of an ITS application through a well-defined organisational and operational framework with the aim of contributing to user safety, efficiency, comfort and/or to facilitate or support transport and travel operations. ITS service provider means any provider of and ITS service, whether public or private. ITS user is any user of ITS applications or services including travellers, vulnerable road users, road transport infrastructure users and operators, fleet managers and operators of emergency services.

ITS integrate telecommunications, electronics and information technologies with transport engineering in order to plan, design, operate, maintain and manage transport systems. The application of information and communication technologies to the road transport sector and its interfaces with other modes of transport will make a significant contribution to improving environmental performance, efficiency, including energy efficiency, safety and security of road transport, including the transport of dangerous goods, public security and passenger and freight mobility, whilst at the same time ensuring the functioning of the internal market as well as increased levels of competitiveness and employment.

The conclusion from many years of research conducted in the USA and Canada is that, the use of ITS results in the reduction of the funds allocated for the transport infrastructure even by 30 – 35 %, with the same functionality of the system (FHWA-OP-03-XXX 2005).

1.2 Development phases of ITS

Based on the analysis of the literature, it is possible to select three phases in the history Intelligent Transport Systems development to date – fig. 2.

First phase is the beginning of ITS research in the 1970 and 1980s. Since the 1970's, several European companies have developed more complex systems that broadcasted a code at the start of the message so that only cars affected by that information would receive it. In Germany, ARI, a highway radio system using FM (Frequency Modulation), was introduced in 1974 to alleviate traffic congestion on northbound autobahns during summer holidays.

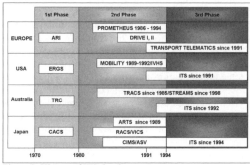

Figure 2. History of ITS development

Legend:
- ARI (Auto-fahrer Rundfunk Information),
- ERGS (Electronic Route Guidance System),
- TRC (Traffic responsive Capabilities),
- CACS (Comprehensive Automobile Control System,
- ARTS (Adaptive Responsive Traffic System),
- TRACS (Traffic Responsive Adaptive Control System),

- RACS (Road/Automobile Communication System,
- VICS (Vehicle Intelligent Control System),
- CIMS (Control Intelligent Management System),
- ASV (Advanced System of Vehicle).

Since 1970, the Department of Main Roads in Australia installed the first system that included 30 signalized intersections featuring centralized control and TRC.

In the United States, government sponsored in-vehicle navigation and route guidance system - ERGS was the initial stage of a larger research and development effort called the ITS (Dingus & 1996). In 1973 the Ministry of International Trade and Industry (MITI) in Japan funded the Comprehensive Automobile Control System (CACS) (Dingus 1996 & Tokuyama 1996). All of these systems shared a common emphasis on route guidance and were based on central processing systems with huge central computers and communications systems. Due to limitations, these systems never resulted in practical application.

In the *second phase* from 1981 and 1994 the conditions for ITS development were determined. Technological reforms, such as the advent of mass memory, made information processing cheaper. New research and development efforts directed at practical use got under way. Two projects were being run in Europe at the same time: the Program for a European Traffic System with Higher Efficiency and Unprecedented Safety (PROMETHEUS), which was mainly set up by auto manufacturers, and the Dedicated Road Infrastructure for Vehicle Safety in Europe (DRIVE), set up by the European Community. PROMETHEUS was started in 1986 and was initiated as part of the EUREKA program, a pan-European initiative aimed at improving the competitive strength of Europe by stimulating development in such areas as information technology, telecommunications, robotics, and transport technology. The project is led by 18 European automobile companies, state authorities, and over 40 research institutions.

In 1991 ERTICO (European Road Transport Telematics Implementation Coordination Organization) was created with support of EC as a private-public partnership, and is open to all European organizations or international organizations operating substantially in Europe with an interest in ITS. It facilitates the safe, secure, clean, efficient and comfortable mobility of people and goods in Europe through the widespread deployment of ITS. Specifically, ERTICO:
- provides a platform for its Partners to define ITS development & deployment needs,
- acquires and manages publicly funded ITS development and deployment projects on behalf of its Partners,
- formulates and communicates the necessary European framework conditions for the deployment of ITS,
- Enhances the awareness of ITS benefits amongst decision makers and opinion leaders.

Applied effectively, ITS can save lives, time and money as well as reduce the impact of mobility on the environment. ERTICO's vision is of a European transport system that is safer, more efficient, and more sustainable and more secure than today. ITS technology, combined with the appropriate investment in infrastructure, will have reduced congestion and accidents while making transport networks more secure and reducing their impact on the environment.

In Japan, work on the RACS project, which formed the basis for current car navigation system, began in 1984. In 1985, a second generation traffic management system was installed in Australia. This was known as the TRACS.

In 1989 in the USA the Mobility 2000 group was formed and led to the formation of IVHS (Intelligent Vehicle Highway Systems) America in 1990, whose function was to act as a Federal Advisory Committee for the US Department of Transportation. IVHS program was defined as an integral part, became law in order to develop "a national intermodal transport system that is economically sound, to provide the foundation for the nation to compete in the global economy, and to move people and goods in an energy-efficient manner".

In 1991 ITS America was established as a not-for-profit organization to foster the use of advanced technologies in surface transportation systems. Members include private corporations, public agencies, academic institutions and research centres. The common goal is to improve the safety, security and efficiency of the U.S. transportation system via ITS. Traffic accidents and congestion take a heavy toll in lives, lost productivity, and wasted energy. ITS enables people and goods to move more safely and efficiently through a state-of-the-art, multi-modal transportation system. ITS America has sister organizations in Europe and Japan, as well as affiliates in Canada, Brazil, and elsewhere.

The *third phase* began in 1994, when the practical applications of earlier programs were seen, understood, and intelligent transportation systems were being thought of in intermodal terms rather than simply in terms of automobile traffic. ITS have started to gain recognition as critical elements in the national and international overall information technology hierarchy.

In 1994 the IVHS program (USA) was renamed the ITS (Intelligent Transportation Systems) indicating that besides car traffic also other modes of transportation receive attention and during the first world congress in Paris, the term - Intelligent Transport Systems (ITS) was accepted.

Development of the transport telematics and ITS applications was envisaged in the IV EU Framework Program (1994-1998). The 4th Framework Program adopted by the Council and Parliament in April 1994 includes telematics as a major topic of research. It invites the Commission to draw up Telematics Applications for Transport in Europe Program (4 November 1994) for the measures required at Community level for the implementation of Telematics in the Transport Sector (action plan); and to support the work of standardization in traffic management by means of all suitable measures including research and development.

ITS Japan established in 1994 promotes research, development and implementation of ITS in cooperation with five related national ministries in Japan and serves as the primary contact for ITS-related activities throughout the Asia Pacific region. ITS Japan is Part of a Global Advanced Information and Telecommunications Society. The policies of ITS include development of system architecture, research and development (R&D), standardization and international cooperation, and so on. The Interministerial Council works in cooperation with the national and international organizations - such as the Vehicle, Road, and Traffic Intelligence Society (VERTIS) - and supports a variety of activities. VICS (Vehicle Information and Communication System) and ATIS (Advanced Traffic Information System) have been recently in operation in Japan. VICS started from April 1996 in Tokyo and Osaka by VICS Centre supported by Ministry of Construction, Ministry of Telecommunications and National Police Agency and expanding the service area. VICS Centre receives real time traffic information from Highway Traffic Information Centre which gathers the information from each of the highway authorities. And VICS Centre provides the information through roadside beacons as well as FM broadcasting.

In Australia in 1998, the TRAC and South East Freeway's systems merged to create STREAMS Version 1. Since 2007 STREAMS Version 3 was implemented. It is Integrated Intelligent Transport System that provides traffic signal management, incident management, motorway management, vehicle priority, traveller information and parking guidance.

ITSS (Intelligent Transportation Systems Society) is governed in accordance with the Constitution and Bylaws of the Institute of Electrical and Electronics Engineers (IEEE), the basis of ITSS (Press Release announcing the new ITS Council) were implemented in 1999. The purposes of the Society are to bring together the community of scientists and engineers who are involved in the field of interest stated herein, and to advance the professional standing of the Members and Affiliates.

New development of the Intelligent Transport Systems is opened by the program of an EU common transport policy for the years 2001–2010. Addi-

tionally, the European Commission has begun the negotiations, in order to achieve consensus on the introduction in 2010 of an e-Call emergency system in all new cars (the new deadline is 2014).

The matter of transport telematics appeared in Polish publications in the middle of the nineties. In 1997 the attempt was made to define conceptual scope and the area of transport telematics applications (Wawrzyński 2003), which were finally described as a branch of knowledge and technical activities integrating information technology with telecommunication in the applications for the needs of the transport systems.

On the 19 of March 2007 in the district court of Katowice, the registration took place of the Polish Association of Transport Telematics (PATT). It is a newly called gathering, which members dwelling from various environments such like colleges, research institutes, national and private companies of transport business, put themselves for target, through activity in Association, propagating transport telematics and its applications into possible diverse circles of recipients.

31 May 2007 was signed the agreement between PATT and Intelligent Transportation Systems Slovakia, concerning the realization of bilateral contacts and the mutual partnership for the development intelligent transport systems in signatory's' countries.

In 2008 PATT became the Member of the ERTICO – ITS Europe-hosted Network of National ITS Associations.

On the 26 of April 2007 the founder's meeting took place of an Intelligent Transport Systems Association - ITS Poland. The association's objective is to form a partnership of knowledge for the promotion of the ITS solutions, as a means to improving transport efficiency and safety, with the natural environment protection in mind. ITS Poland cooperates with similar organizations in Europe and world wide.

2 STANDARDIZATION OF ITS AREA

2.1 Standardization of ITS

European Intelligent Transport Systems have been fully exploited to maximize the potential of the transport network. European standards will become a key element of the preferred solutions in emerging economies.

Public transport users will have access to up-to-the-minute information, as well as the benefit of smart and seamless ticketing. Freight operators will have real-time information about the entire logistics chain, enabling them to choose the most secure and efficient route for their consignments.

Standardization in transport telematics in Europe is dealt with by the following institutions

(Wawrzyński 2003 & Wydro 2001): CEN, ETSI and CENELEC.

CEN (European Standardization Committee) - is a private technical association of a „non-profit" type, operating within a Belgian legislation, with a seat in Brussels. Officially it was formed in 1974, but the beginnings of its activities date back to – Paris, 1961. The primary task of CEN is drafting, acceptance and dissemination of the European standards and other standardizing documents in all the spheres of the economy, except electro-technology, electronics and telecommunication. Currently CEN has 30 state members. Polish Standardization Committee (PKN) gained the status of a full CEN member on the 1 January 2004.

ETSI – European Institute for the Telecommunication Standards – was formed on the 29 of March 1988, and is the European equivalent of IEEE. The prime objective of ETSI is drafting standards necessary for creation of the European telecommunication market. In 1995 the work of the organization was made international by admitting also the institutions from outside Europe, to participate in it.

CENELEC – European Committee for Electro technical Standardization - was formed in 1973. In Poland the role of the State Committee is performed by Polish Standardization Committee – PKN (it is a CENELEC member since 1 of January 2004).

CENELEC, together with CEN and ETSI form European technical standardizing system, whilst international standards come under the jurisdiction of the International Organization for Standardization (ISO) and International Electro technical Commission (IEC).

In 1991, the Technical Committee for Transport Telematics and Road Traffic - CEN/TC 278 (Road Transport and Traffic Telematics) was established.

Also, a world organization – Telecommunication Industry Association has been established, within which, the Technical Committee ISO/TC 204 is responsible for standardization in Transport Telematics (Intelligent Transport Systems).

In the Committee TC 278, as well as in TC 204, there are working groups, which are responsible for various areas of activities – table 1.

Table 1. Areas of activities for TC 278 and TC 204 working groups

The activity area	TC 278	TC 204
EFC –Electronic fee collection and access control	WG 1	WG 5
FFMS – Freight and Fleet Management systems	WG 2	WG 7
PT – Public Transport	WG 3	WG 8
TTI – Traffic & Traveller Information	WG 4	WG 10
TC – Traffic Control	WG 5	WG 9
GRD – Geographic road data	WG 7	

RTD – Road Traffic Data	WG 8	
DSRC – Dedicated Short Range Communication	WG 9	WG 15
HMI – Human-machine Interfaces	WG 10	
Automatic Vehicle Identification and Automatic Equipment Identification	WG 12	WG 4
Architecture and terminology	WG 13	WG 1
After theft systems for the recovery of stolen vehicles	WG 14	
Safety	WG 15	
Data base technology		WG3
Navigation systems		WG 11
Vehicle/road way warning and control systems		WG 14
Wide area communications/protocols and interfaces		WG 16
Intermodal aspects using mobile devices for ITS		WG 17

TC 278 Technical Committee formulated following standards for the transport telematics: EN 12253, EN 12795, and EN 12834 (ISO 15628) and EN 13372 – table 2.

Table 2. Standards for the transport telematics formulated by TC 278

Standard	Characterization
EN 12253 (2003)	RTTT. DSRC. Physical layer using microwave at 5.8 GHz. Traffic control, Physical layer (OSI), Open systems interconnection, Microwave links, Radio links, Information exchange, Data transmission, Communication networks, Mobile communication systems, Telecommunication systems, Data processing.
EN 12795 (2003)	RTTT. DSRC data link layer. Medium access and logical link control.
EN 12834 (2003)	RTTT. DSRC application layer.
EN 13372 (2003	RTTT. DSRC. Profiles for RTTT applications.

ETSI - European Institute for the Telecommunication standards developed standards EN 300674 and EN 301091, concerning transport telematics – table 3.

Table 3. Standards for the transport telematics developed by ETSI

Standard	Characterisation
ETSI EN 300 674-1 V1.2.1	Electromagnetic compatibility and Radio spectrum Matters (ERM); RTTT; DSRC transmission equipment (500 Kbit/s / 250 Kbit/s) operating in the 5, 8 GHz Industrial, Scientific and Medical (ISM) band; Part 1: General characteristics and test methods for Road Side Units (RSU) and On-Board Units (OBU).
ETSI EN 300	Part 2.1: Harmonized EN under article 3.2

674-2-1 V1.1.1	of the R&TTE Directive; Sub-part 1: Requirements for the Road Side Unit (RSU).
ETSI EN 300 674-2-2 V1.1.1	Part 2.2: Harmonized EN under article 3.2 of the R&TTE Directive; Sub-part 2: Requirements for the On-Board Unit (OBU).

2.2 Standardization of Maritime Intelligent Systems

Maritime telematics applications support routine maritime operations, including navigation, as well as safety purposes.

Maritime intelligent systems involve the use of GPS technologies, wireless mobile communication systems, internet access, which provide vessel tracking, emergency aid and electronic mapping to monitor and provide important boat data from port, land or sea. Systems normally consist of a user interface, satellite antenna, and a communication link with the vessel's electronic systems. This technology can be vital to the user since it provides a satellite link to the outside world when other communications may unavailable. The standards of maritime telematics were presented in table 4.

Table 4. Standards for the maritime telematics by CEN

Standard	Characterization
EN 300065	Narrow-band direct-printing telegraph equipment for receiving meteorological or navigational information (NAVTEX). Part 1: Technical characteristics and methods of measurement. Part 2: Harmonized EN covering the essential requirements of article 3.2. Part 3: Harmonized EN covering the essential requirements of article 3.3.
EN 300066	Float-free maritime satellite Emergency Position Indicating Radio Beacons (EPIRBs) operating in the 406,0 MHz to 406,1 MHz frequency band. Technical characteristics.
EN 300162 -1	Radiotelephone transmitters and receivers for the maritime mobile service operating in VHF bands. Part 1: Technical characteristics and methods of measurement.
EN 300225	Technical characteristics and methods of measurement for survival craft portable VHF radiotelephone apparatus.
EN 300338	Technical characteristics and methods of measurement for equipment for generation, transmission and reception of Digital Selective Calling (DSC) in the maritime MF, MF/HF and/or VHF mobile service,
EN 300373 -1	Maritime mobile transmitters and receivers for use in the MF and HF bands; Part 1: Technical characteristics and methods of measurement.
EN 300698 -1	Radio telephone transmitters and receivers for the maritime mobile service operating in the VHF bands used on inland waterways; Part 1: Technical characteristics and methods.
EN 300720 -1	Ultra-High Frequency (UHF) on-board communications systems and equipment; Part 1: Technical characteristics and methods of measurement.
EN 301025 -1	VHF radiotelephone equipment for general communications and associated equipment for Class 'D' Digital Selective Calling (DSC); Part 1: Technical characteristics and meas.
EN 301033	Technical characteristics and methods of measurement for reception of ship borne watch keeping receivers for reception of DSC in the maritime MF, MF/HF and VHF bands.
EN 301178 -1	Portable Very High Frequency (VHF) radiotelephone equipment for the maritime mobile service operating in the VHF bands (for non-GMDSS applications only); Part 1: Technical characteristics and methods of measurement.
EN 301403	Maritime Mobile Earth Stations (MMES) operating in the 1,5 GHz and 1,6 GHz bands providing voice and direct printing for the Global Maritime Distress and Safety System (GMDSS); Technical characteristics and methods of measurement.
EN 301466	Technical characteristics and methods of measurement for two-way VHF radiotelephone apparatus for fixed installation in survival draft.
EN 301688	Technical characteristics and methods of measurement for fixed and portable VHF equipment operating on 121,5 MHz and 123,1 MHz.
EN 301843 -1	Electromagnetic Compatibility (EMC) standard for marine radio equipment and services; Part 1: Common technical requirements.
EN 301925	Radiotelephone transmitters and receivers for the maritime mobile service operating in VHF bands. Technical characteristics and methods of measurement.
EN 301929 -1	VHF transmitters and receivers as Coast Stations for GMDSS and other applications in the maritime mobile service. Part 1: Technical characteristics and methods.
EN 302152 -1	Satellite Personal Locator Beacons (PLBs) operating in the 406, 0 MHz to 406, 1 MHz frequency band; Part 1: Technical characteristics and methods of measurement.
EN 302194 -1	Navigation radar used on inland waterways: Part 1: Technical characteristics and methods of measurement.
EN 302752	Active radar target enhancers; Harmonized EN covering the essential requirements of article 3.2 of the R&TTE Directive.

2.3 Actual activities of ITS standardization in EU

ITS standards define how ITS systems, products, and components can interconnect, exchange information and interact to deliver services within a transportation network. ITS standards are open-interface standards that establish communication rules for how ITS devices can perform, how they can connect, and how they can exchange data in order to interoperate. It is important to note that ITS standards are not design standards: They do not specify specific products or designs to use. Instead, the use of standards gives transportation agencies confidence that components from different manufacturers will work together, without removing the incentive for designers and manufacturers to compete to provide products that are more efficient or offer more features.

The ability of different ITS devices and components to exchange and interpret data directly through a common communications interface, and to use the exchanged data to operate together effectively, is called *interoperability*. Interoperability is key to achieving the full potential of ITS. Seamless data exchange would allow an emergency services vehicle to notify a traffic management center to trigger change in the timing of the traffic signals on the path to a hospital, in order to assist the responding ambulance.

Interoperability is defined as the ability of ITS systems to:
- Provide information and services to other systems
- Use exchanged information and services to operate together effectively.

The European Commission Mandate M/453 invites the European Standardisation Organisations - ESOs (ETSI, CEN, CENELEC), to prepare a coherent set of standards, technical specifications and technical reports within the timescale required in the Mandate to support European Community wide implementation and deployment of interoperable Co-operative Intelligent Transport Systems (Co-operative ITS).

Intelligent Transport Systems (ITS) means applying Information and Communication Technologies (ICT) to the transport sector (M/453). ITS can create clear benefits in terms of transport efficiency, sustainability, safety and security, whilst contributing to the EU Internal Market and competitiveness objectives. To take full advantage of the benefits that ICT based systems and applications can bring to the transport sector it is necessary to ensure interoperability among the different systems throughout Europe at least.

This Mandate supports the development of technical standards and specifications for Intelligent Transport Systems (ITS) within the European Standards Organisations in order to ensure the deployment and interoperability of Co-operative systems, in particular those operating in the 5 GHz frequency band, within the European Community. Standardisation is a priority area for the European Commission in the ITS Action Plan in order to achieve European and global ITS co-operation and coordination.

Standardisation for Cooperative ITS systems has already been initiated both by ETSI and ISO as well as within other international standards organisations. European standardisation activities to provide standardised solutions for Cooperative ITS services are therefore closely related to the world wide standardisation activities.

Within three months of the date of acceptance of this Mandate ETSI, CEN and CENELEC must present a report to the Commission with the work program to achieve goal of completion of the standardization process for Cooperative ITS services.

Particular attention must be given to the involvement of all relevant parties, including public authorities, and to the working arrangements between relevant industry forums and consortia.

Within one year of the date of acceptance of this Mandate ETSI, CEN and CENELEC must present a progress report on the achievements in accordance with the work program. CEN, CENELEC and ETSI must present annual progress reports to the Commission services.

Twenty months after the acceptance of this mandate, a comprehensive report must be presented with the status of the on-going work and the latest available draft of the different standards.

The European Commission mandate on Cooperative Intelligent Transport Systems requires the synchronization among the European Standards Organizations on one hand; on the other hand it recommends collecting feedback from stakeholders affected by that standardization work. This session intends to verify if all the bits and bytes of standardization fit to each other, to identify shortcomings and potential show-stoppers and to find proposals for challenging standardization issues. In addition, the session offers the possibilities to present topics that should be considered by standardization additionally.

3 CONCLUSIONS

Intelligent Transport Systems are integral part of European Transport Policy. ITS Directive is the legal instrument for the deployment of ITS in Europe. Standardisation has a major role in the development of interoperable ITS. Interoperability and building ITS architecture brings about the necessity to develop standards concerning, among the others, technical, safety solutions as well as data transmission protocols between the system elements and it environment solutions. These applications in the future may provide quick and precise information and allow to safely managing transport. In the forthcoming years they will be further improved by using Galileo system, whose localizing precision will be better than that of GPS. Integration of tools by using standards would allow: reducing times and errors (preventing re-typing), facilitating engineering & trading, improving data recording, improving survey, maintenance and repair (life cycle). Telematics is a vital means of development for maritime transport in the European Union.

One of the key benefits of ITS is the exchange of information and completion of transactions directly between computers, eliminating the need for processing purchase orders, bills of lading or invoices. Clear, constructive, harmonised, and easy applicable legal rules affect differently the economic parameters of maritime transport than vague and contradictory legal rules or even more the absence of legal provisions. Community legislation now exists for all modes of transport creating new open market conditions.

The European Commission Mandate M/453 on Cooperative Intelligent Transport Systems was approved by CEN and ETSI.

Furthermore, within the frame of high level agreements between the European Union, US Department of Transportation and the Japanese communication ministries on global activities to harmonize standardization and cooperative ITS applications as well as a roadmap for deployment, this high level managers round table will provide the latest news on the global activities and discuss the way forward to achieve global interpretability for cooperative ITS when implemented and deployed in a few years.

In September 2010 the standard ETSI EN 302 665 specifying the ITS Communications Architecture has been published. Although the architecture has been designed in a modular way that allows flexible usage and implementation it is still required to harmonize the internal interfaces between the modules and the interfaces to the external world.

REFERENCES

Berghout, L. & Bossom, R. & Chevreuil, M. & Burkert A. & Franco, G. & Gaillet, J. F. & Pencole, B. & Schulz, H. J. 1999. Transport Telematics System Architecture. Constraint analysis, mitigation strategies and recommendations. Directorate-General XIII of European Commission on Information Society. Brussels.

Dingus, T. & Hulse M. & Jahns S. & Alves-Foss, J. & Confer, S. & Rice, A. & Roberts, I. & Hanowski, R. & Sorenson D. 1996. Development of Human Factors Guidelines for Advanced Traveller Information Systems and Commercial Vehicle Operations. Literature Review. November.

Directive 2000/40/EC of the European Parliament and of the Council of 7 July 2010 on the framework for the deployment of Intelligent Transport Systems in the field of road transport and for interfaces with other modes of transport. EN Official Journal of the European Communities L 207.

Electronic Toll Collection/Electronic Screening Interoperability Pilot Project Final Report Synthesis. 2005. Department of Transportation, Publication FHWA-OP-03-XXX, USA, July 29.

Goel, A. 2007. Fleet Telematics – Real – Time management and Planning of Commercial Vehicle Operations. Operations Research. Computer Science Interfaces Series. Vol. 40. Springer.

Möglichkeiten und Grenzen des Einsatzes von Telematik im Verkehr. 2003. Internationales Verkehrswesen. Nr 12. Hamburg, s. 599-607.

Mikulski, J. 2007. Chair of Automatic Control in Transport, Faculty of Transport. Advances in Transport Systems Telematics. Silesian University of Technology, Katowice.

M/453 EN. Standardisation mandate addressed to CEN, CENELEC and ETSI in the field of information and communication technologies to support the interoperability of co-operative systems for Intelligent Transport in the European Community. European Commission Enterprise and Industry Directorate-General Innovation Policy. ICT for Competitiveness and Innovation Brussels, 6th October 2009. DG ENTR/D4.

Nora, S. & Minc, A. 1978. L'Informatisation de la société. Rapport à M. le Président de la République. La Documentation Française, Paris. English version. 1980. The Computerization of Society. A report to the President of France. MIT Press, Cambridge, Massachusetts.

Nowacki, G. 2008. Road transport telematics. Monograph. Motor Transport Institute, Warsaw.

Piecha, J. 2003. Register and data process in transport telematics systems. Monograph, Silesian University of Technology, Gliwice.

Tokuyama, H. 1996. Intelligent transportation systems in Japan. Public Roads.

Wawrzyński, W. 2003. Telematics place in science discipline of transport. Transport Faculty of Warsaw University of Technology, Warsaw.

Wydro, K. 2001. Normalization in transport telematics. Telecommunication and Information Techniques, No 3-4, Warsaw.

Wydro, K. 2005. Telematics – meaning and definition. Telecommunications and information techniques. No 1-2. Communications Institute, Warsaw.

6. Computer Systems Aided Management in Logistics

K. Chwesiuk

Maritime University of Szczecin, Szczecin, Poland

ABSTRACT: This paper aims at presenting a concept of an integrated computer system of management in logistics, particularly in supply and distribution chains. Consequently, the paper includes the basic idea of the concept of computer-based management in logistics and components of the system, such as CAM and CIM systems in production processes, and management systems for storage, materials flow, and for managing transport, forwarding and logistics companies. The platform which integrates computer-aided management systems is that of electronic data interchange.

1 INTRODUCTION

Taking into account the development of computer technologies, we can classify today's production processes as follows:
1 independent, computer controlled machining and assembly stations (CM – Computer Module),
2 FMS – Flexible Manufacturing Systems,
3 CAM – Computer Aided Manufacturing Systems,
4 CIM – Computer Integrated Manufacturing Systems.

Typical operations in today's production systems include technological (machining and assembly), control, transport, storage operations and their combinations. Besides, there are processes of component and raw material supply, co-operation, distribution of finished products and after sale services.

Logistics come to assistance in managing the production system understood in such broad terms. There are clearly distinguished areas of logistics:
– material supply,
– co-operation,
– production,
– distribution.

Processes taking place in these four areas of logistics require efficient management. To improve the system of logistic management of production we have to design and implement a computer system.

This paper presents the idea of a wide range computer system which aids the management of production system logistics.

2 THE CONCEPT OF COMPREHENSIVE COMPUTER SYSTEM OF MANAGEMENT IN LOGISTICS

The production system consists of four subsystems:
1 materials supply, handled by materials supply logistics,
2 co-operation, handled by co-operation logistics,
3 manufacturing, handled by manufacturing logistics,
4 finished goods distribution, handled by distribution logistics.

Figure 1 graphically illustrates a logistic chain of materials supply for the manufacturing process in a production company. Participants of this chain are as follows:
– original suppliers,
– suppliers of components and subassemblies,
– supply centers (see Fig. 1).

Figure 2 shows a logistic chain of co-operation in the manufacturing process in a production company. There are two types of business entities in this chain:
– suppliers to co-operators,
– co-operators (see Figure 2).

Figure 3, in turn, presents graphically a logistic chain of distribution of finished goods from one particular manufacturer. This chain comprises such business entities as:
– distribution centers,
– wholesale and retail stores,
– end recipients (see Figure 3).

Figure 4 illustrates graphically the concept of a full-range computer-based management system in

the production process logistics. Its component systems are as follows:
- computer-aided manufacturing – CAM,
- computer integrated manufacturing – CIM,
- material requirement planning (MRPI) and manufacturing resource planning (MRPII),
- management of finished goods distribution – SD (see Figure 4).

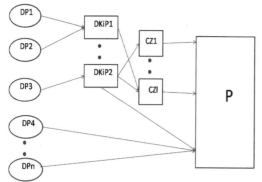

Figure 1. Logistic chain of materials supply. DP – original supplier, DKiP – supplier of components and subassemblies, CZ – distribution centre, P – producer. Source: author's study based on.[1]

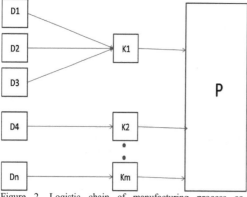

Figure 2. Logistic chain of manufacturing process co-operation. D – supplier to a co-operator, K – co-operator in manufacturing process, P - producer. Source: author's study based on.[2]

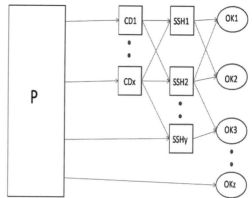

Figure 3. Logistic chain of finished goods distribution. CD – distribution centre, SSH – wholesale and retail network, OK – end recipient, P - producer. Source: author's study based on.[3]

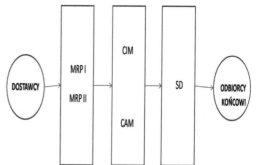

Figure 4. Concept of computer integrated management system in production logistics. MRPI – computer system for material requirement planning, MRPII – manufacturing resource planning system, CAM – computer-aided manufacturing system, CIM – computer integrated manufacturing system, SD – computer-aided goods distribution management system. Source: author's study

The chart of a computer integrated management system in production logistics shown in Fig. 4 indicates with arrows the direction of material flow, or to be exact, the flow of all production factors involved in the process of manufacturing a finished product and its distribution to end users. However, electronic flows of data between the main components of the comprehensive computer system run in the opposite direction. It is in the first step of the logistic distribution chain that information on the demand for given maker's products is recognized and processed. This is done in the computer system of distribution by collecting and aggregating orders from end recipients for given product models and types of a given manufacturer. Besides, in projecting the product demand the amounts in stock of each member of the logistic distribution chain are taken into account. Data from the computer-based distribu-

---

[1] See Śliwczyński B.: Planowanie logistyczne. Podręcznik do kształcenia w zawodzie technik logistyk. Biblioteka Logistyka, Poznań 2007
[2] See. Śliwczyński B.,: Planowanie logistyczne ….

[3] See Śliwczyński B.: Planowanie logistyczne….

tion system are transferred to production management systems, i.e. CAM and/or CIM, which are operated in the production company.

After information is processed in CAM and/or CIM systems, the latter in particular, the resultant information is obtained in the form of, e.g. a plan and schedule of manufacturing. This information and the data from current monitoring of the materials in stock and the state of materials flows in production lines are directed to MRPI and/or MRPII systems.

Based on the computer-aided systems of production resources control the material flow is managed in the logistic chain of materials supply and the logistic chain of production co-operation, from original suppliers and co-operators to producers.

The computer integrated management in production logistics comprises several computer systems, used in such areas as:
– forwarding,
– transport services,
– transport terminal services,
– customs offices,
– banks,
– insurance companies
– standardization offices,
– others.

The key condition for successful design and operation of a computer integrated management in production logistics is that the co-operating computer systems share the relevant information. This objective is obtained by access to a common integrated data base (data warehouse) and by the use of common standards of electronic data interchange – EDI.

3 BRIEF CHARACTERISTICS OF BASIC COMPONENTS OF THE COMPUTER INTEGRATED SYSTEM OF MANAGEMENT IN LOGISTICS

3.1 *Computer Integrated Manufacturing – CIM*

The basic aim of CIM is a comprehensive computer-aided system for integrated implementation of production orders. One can say it is an integrated system of production order execution.

All data that appear in manufacturing processes from material supply, through work engineering to manufacturing and assembly, should always be utilized in planning tasks. While planning production operations the planner should send all data to the production area through strictly defined information channels. All data connected with manufacturing and executed orders are stored in the central data base. Production data are created mainly during the design of a product, then data from orders are added in the planning phase. These data are crucial for manufacturing and assembly. Data that appear in the above areas, i.e. dates/times of completing each

manufacturing operation or operational loads of each machine, device and work station are included in the central data base and can be used by the planning system provided that data from the actual manufacturing and assembly units are sent back via a company's data base system or DCN – Direct Numerical Control system.[4]

As computer technologies develop, better technical conditions are being created for the construction of more advanced production facilities that may run automatically, with limited participation of people. The role of the human in such systems focuses on issues such as the programming of computers and computer-controlled production equipment.[5]

The growing presence of computer systems in all spheres of manufacturing company operations and integration of these systems into one all-encompassing computer system brings about many changes in technological processes. Some of these changes lead to:
– shortened time of preparing and executing production orders,
– reduction of operating costs,
– improved internal and external communication,
– more effective design, planning and preparation of production.[6]

One aspect worth emphasizing is increased utilization of company's production capacity by using the company's data base, which allows to eliminate doing the same work twice. Besides, errors due to insufficient communication are avoided. The integration of computer systems enhances the flexibility of production processes, particularly manufacturing processes, thus the manufacturer is able to respond faster to customer's requests, which often refer to details of one particular order.[7]

The CIM system consists of two interconnected subsystems:
1 CAD – Computer Aided Design, which is composed of the mutually co-operating subsystems:
  – CAE – Computer Aided Engineering; its task is to design and engineer new products or to modernize products already made,
  – CAP – Computer Aided Planning and CAPP – Computer Aided Process Planning; these are supposed to prepare the production process in terms of technology, i.e. product construction, technology of manufacturing product parts, subassemblies and the finished product, preparing technical drawings, lists of components and the organization of the machining and assembly process,

---

[4] See Durlik I.: Inżynieria Zarządzania, Strategia i projektowanie systemów produkcyjnych, Cz. I. Agencja Wydawnicza „PLACET", Warszawa 1998
[5] See Durlik I.: Inżynieria Zarządzania ….
[6] See Durlik I.: Inżynieria Zarządzania ….
[7] See Durlik I.: Inżynieria Zarządzania ….

- CAD – Computer Aided Design, whose task is to plan the operation of the manufacturing system comprising the manufacturing of parts, assembly of components and the whole product, including measurements, packaging and dispatch of finished products,
- data base (DB) and expert systems (ES), which enable the functioning of all the areas of computer-aided production together with a expert knowledge base (KB) co-operating with these systems;
2 A subsystem of Computer Aided Manufacturing (CAM), which will be described in the next chapter.

3.2 *The system of Computer Aided Manufacturing – CAM*

The Computer Aided Manufacturing (CAM) is defined as a system for preparing programs for the process of manufacturing, control and recording data on the manufacturing output. This system also encompasses such organization functions as production planning, setting the dates of materials and subassembly supply from co-operators or the delivery of finished products.

CAM can be described as:
- a flexible manufacturing system, which is capable of manufacturing at the same time sets of various products of different series size, where quantities and assortments are changed by a computer,
- hierarchically controlled system; computer-supervised and handled by a small team, making up less than 10 percent of the company personnel that would be necessary to perform the same tasks in conventional conditions.[8]

The system which generates software for the machining and paths along which parts and subassemblies will pass through work modules and stations, while these programs and paths are optimized relative to work load and the degree of utilization of machines and assembly devices, production cycles, productivity, energy consumption, environment pollution and work security.[9]

In industrial practice CAM systems, apart from the manufacturing in flexible production systems, also include:
- development of software, or operating plans of machining and assembly, that as a rule are variable depending on the current production situation,
- planning of component paths and schedules of the production,
- optimal manufacturing control,
- optimal product quality control,
- production management.[10]

The computer-aided manufacturing – CAM – is regarded as a development of the designed and functioning flexible manufacturing systems with some functions connected with control at a level of a specific production system. The CAM system is often treated as a transitory stage leading to the computer integrated manufacturing (CIM).

The CAM system consists of the following subsystems:
1 CAMC – Computer Aided Manufacturing Control; its basic function is programming and computer-aided control of numerically-controlled manufacturing equipment,
2 CAQ/CAQC – Computer Aided Quality/Computer Aided Quality Control; this subsystem is designed to provide the highest standard of product quality,
3 CAT – Computer Aided Testing, for examining the technical condition of machines and tools.

A production company using computer-aided manufacturing should have the following technological machines and facilities:
- numerically controlled (NC) machines tools,
- machine tools with CNC (Computer Numerical Control),
- machine tools with DNC (Direct Numerical Control),
- IR – Industrial Robots,
- IM – Industrial Manipulators,
- AS – Automated Storage,
- AGV – Automated Guided Vehicles.

The use of CIM and CAM systems requires specific input data, such as production execution orders and data on future demand for the products offered. These data are acquired from the computer system handling distribution logistics. Output data, on the other hand, after processing in the CIM and CAM systems, are production schedules, which themselves constitute input data for computer systems of materials supply and co-operation.

3.3 *Enterprise Resource Planning - ERP*

The computer system of an ERP class can be defined as a set of integrated functional modules, optimizing internal and external business processes, those occurring in the immediate environment of the enterprise. Such optimization is possible through the offering of ready tools enabling automation of data exchange with co-operators within the entire logistic chain. The **main features** of the ERP computer system can be set forth as follows:
- functional complexity – includes all spheres of technical and economic activities of an enterprise; it is implemented within the company functional structure,

---

[8] See Durlik I.: Inżynieria Zarządzania ....
[9] See Durlik I.: Inżynieria Zarządzania ....
[10] See Durlik I.: Inżynieria Zarządzania ....

- integration of data and processes – refers to data exchange inside an object (between the modules) and with the environment (e.g. through an EDI – Electronic Data Interchange); this feature is implemented within the information structure,
- structural and functional flexibility – ensures maximum adjustment of hardware-software solutions (implemented within the technical and functional structures) to suit the needs of an object at the moment the system is installed and started up; it also enables its dynamic adjustment when the environment generates variable requirements and needs,
- openess – assures the ability to extend the system with new modules, scalable architecture (usually customer -server) and creation of links with external systems, e.g. systems of market partners,
- substantial advancement – ensures full computer aided support of information-decision processes, using mechanisms of free data extraction and aggregation, seeking variants, optimization, projecting etc., as well as, in practice, basing the system on, *inter alia*, such concepts of logistic management as delivery *Just in Time (JiT)*, production control according to MRP II standards (*Manufacturing Resource Planning*), MRP II Plus (*MRP - Money Resource Planning* - MRP II developed with financial procedures, e.g. cash flow), the ABC method (*ABC - Activity Based Costing*), Total Quality Management and ISO 9000 standards,
- technological advancement – guarantees the compliance with present standards of software and hardware, making it possible for the system to migrate to new platforms of computer equipment, operating systems, communication media and protocols; it offers a graphical interface and use of, generally, relational data base (due to easy way of creating inquiries), with application of fourth generation programming tools etc.,
- conformity with Polish legislation, e.g. with the Act on accounting, in particular the regulations on book-keeping with the use of information technology, principles of reporting the financial performance of a business facility, principles of preparing financial statements etc.[11]

These systems cover all areas of company operation (finance, logistics, production, human resources), optimize internal processes as well as external processes taking place in the near environment of the company, by offering ready tools and allowing to automate data exchange with co-operators in the whole logistic chain. They also have a capability of dynamic configuration, which enables the adjustment of their functionality to the specific operations of an enterprise or other organization.[12]

- The ERP system comprises the following areas of logistic activities:
- customer service – customer data base, order processing, handling individual orders (products on request: *assembly-to-order*, *make-to-order*), electronic data interchange (EDI),
- production – handling of resources, product cost estimation, purchase of raw materials and components, production scheduling, management of product change (introduction of improvements), projection of production capability, determination of critical level of stocks/resources, production process control (e.g. tracking of a product in a manufacturing plant) etc.,
- finance – accounting, control of accounting documents flow, financial settlements, preparation of financial statements as required by the recipient groups (e.g. for the head office and branches),
- integration of the logistic chain – feature that is likely to determine future directions ERP systems will follow, extending their coverage outside the enterprise.[13]

3.4 Computer-aided Supply Chain Management - SCM

SCM class solutions offered on the market are technologically advanced systems. As a rule, they consist of a group of integrated applications serving various areas of logistic chain management. The basic SCM element is material flow planning at each stage, from material extraction to the delivery of ready product to the consumer, through joint product design, demand and supply planning, monitoring stocks level, shipment dispatch organization, joint information management.[14]

The **integrating function** of SCM systems is also their important feature. It is understood as multifunctional integration – enabling integration and optimization of the main enterprise functions at the planning and execution level,
- integration of many enterprises – using Internet capabilities of communications between enterprises and their business partners and customers,
- integration with other systems within the enterprise – enabling convergence of data with transaction systems (including ERP systems, spreadsheets, data bases, text files).[15]

Complex supply chain management is strictly connected with the occurrence of eight mutually supplementing **business processes**. These are:

[11] See Adamczewski P.: Zintegrowane systemy zarządzania ERP/ERPII, Difin, Warszawa, 2003

[12] See Majewski J.; Informatyka dla logistyki, Biblioteka Logistyka, Poznań 2002
[13] See Majewski J.: Informatyka dla logistyki ...
[14] See Długosz J.: Nowoczesne technologie w logistyce, PWE, Warszawa 2009
[15] See Długosz J.: Nowoczesne technologie w logistyce ...

- CRM – Customer Relationship Management. This process enables creating a model supporting optimal building, development and maintenance of contacts with customers. Basically, it identifies market segments, allows to generate criteria for customer grouping, and estimate their profitability. All data generated by a CRM system must be measurable, so that an appropriate cost, sales and investment strategy is developed.
- Customer Service Management. Within this application the customer is able to have a constant access to check product availability, delivery dates or delivery status. Access to current information is guaranteed by an interface connected with manufacturer's production and logistic plans. This module supplements data generated by the CRM with planning procedures which define the method of delivery and product supervision for the customer.
- Demand Management. The main function of demand management is to maintain an optimal balance between customer expectations and production capabilities of the manufacturer. Demand management has advanced projection methods, where projected results are synchronized with the production, purchase and distribution. Besides, this process makes it possible to respond immediately to any internal and external disturbances in the process by generating substitute plans.
- Order Execution. Effective order execution calls for the integration of production, logistic and marketing plans of the manufacturer. The manufacturer should attempt to maintain good relations with suppliers within the supply chain, in order to provide added value to customers and reduce product delivery costs resulting from their geographical location, characteristics of raw materials offered and the selection of transport modes.
- Manufacturing Flow Management. The process is directly related with flexible manufacturing of products, their quality control, analysis of deviations and continuous control of stocks in warehouses. There is a close collaboration of manufacturing flow management module with CRM aimed at building an optimal production infrastructure.
- Supplier Relationship Management (SRM). In a sense, SRM reflects the capabilities of CRM. The difference is, however, that SRM influences product and service suppliers. SRM is supposed to identify and build close business relations with Key Suppliers (classification of suppliers by their profitability, development opportunities and methods of servicing sold products).
- Product Development and Sales. The key importance is attributed to how fast a new product or improved product can be launched on the market; in this way SCM integrates customers and suppliers in the process of product development.

- Claim Management. Effective claim management is a major component of SCM. Many companies neglect this aspect, while it turns out to be an essential factor for the company to gain competitive advantage. The process requires good knowledge of environment protection issues and some legal aspects related with product use procedures.[16]

4 CONCLUSION

The presented concept of integrated computer system of management in logistics makes use of computer-aided systems already employed in management and control of manufacturing processes (CAM and CIM), those used in the logistics of materials supply and co-operation (MRPI, MRPII and ERP) and in distribution logistics (WMS and CMR). The integrated computer system also incorporates computer systems supporting management in forwarding, transport, banking, insurance, customs etc.

The electronic data interchange (EDI) is the platform used for the integration all the above mentioned systems.

REFERENCES

1. Adamczewski P., *Zintegrowane systemy informatyczne w praktyce*, Wydawnictwo MIKOM, Warszawa 2004
2. Długosz J.: *Nowoczesne technologie w logistyce*, PWE, Warszawa 2009
3. Douglas M. Lambert [et al.] ; tł Michał Lipa *Zarządzanie łańcuchem dostaw*, (Tyt.oryg.: *Harvard Business Review on Supply Chain Management*) HELION, Gliwice 2007
4. Durlik I.: Inżynieria Zarządzania, *Strategia i projektowanie systemów produkcyjnych, Cz. I.* Agencja Wydawnicza „PLACET", Warszawa 1998
5. Lech P., *Zintegrowane systemy zarządzania ERP/ERP II*, Difin, Warszawa, 2003
6. Majewski J.: *Informatyka dla logistyki*, Biblioteka Logistyka, Poznań 2002
7. Śliwczyński B.: Planowanie logistyczne. Podręcznik do kształcenia w zawodzie technik logistyk. Biblioteka Logistyka, Poznań 2007

---

[16] See Douglas M., Lambert [et al.]; tł Lipa M.: Zarządzanie łańcuchem dostaw,. HELION, Gliwice 2007

7. Information in Transport Processes

K. B. Wydro
University College of Technology and Business in Warsaw, Poland

ABSTRACT: A paper concerns the problems of information ordering in intelligent transport systems accordingly their role and meaning for various transport processes. There is given an attempt to rules of classification of information, their standardisation questions, reduction of redundancy and false specimens. All these questions are of great importance from the information accessibility, usefulness and reliability point of view. Also are discussed problems of the information selection, it's protection, and it's evaluation as a factors influencing possible improvement of the transport decisions making. Finally, some outline of standardisation rules are presented.

1 INTRODUCTION

Dynamic development of the modern information technologies applied to all the activities in transport, generally called as intelligent transport systems (ITS) becomes possible thanks for growing accessibility to ICT solutions. But really just the broadened accessibility to information becomes crucial for great progress in the discussed area. Information, being beside energy and technical means an indispensable factor for realization of the all transport tasks, may have a decisive meaning for high effectiveness achievement. It is easy to observe, that in the last tens of years, thanks to new technological possibilities appeared not noticed former access to huge amount of ready or possible to obtain information. At the same time, in provided information emerge also considerable content or volume redundant, and even false or erroneous specimens, what obviously reduce effectiveness of the transport activities based on such information and certainly – even all the information processing tasks.

Generally observed growth of information meaning, resulted from progress of ability and possibility of it's intensified usage allowing all human activities improvement, caused even development of the research over this peculiar good, headed toward further improvement of it's usage. Among others there are conducted research on information evaluation, reduction of the redundancy and incidentality, extraction of the valuable parts, mainly these indispensable. It combines with the necessity of developing proper rules for broadly understood information management, what particularly concerns transport branch. Just there exists a need to work out standards (in meaning of mediocre type, pattern, model) and norms, which could be applied in particular areas of information applications in transport. It should improve the effectiveness of operations of various transport systems, level of the cohesion of transport activities and – may be first of all – it's security.

With reference to existing state of ITS resulted by spontaneous and incidental development (Wydro 2006), it should mean ordering of information management and processing according to it's content, what should give possibility to remove it's redundant part and processing of this part, but mainly considerable gains coming from more effective systems operations, (as at actual state systems are weakly co-ordinated or not co-ordinated at all as a result of compatibility lack). Actually even information exchange between systems and equipment made by various producers and providers – often necessary – leads to additional costs and lower reliability of ITS as a whole. It became also a big obstacle for introduction to systems the new functionalities; it's extension or improvement. In particular, emerging in last time tendencies for creation of multi-modal transport structures, rising the security level, providing better transport conditions with information services to millions of individual recipients with very diversified needs profiles, requires efficient systems co-operation an – in some extend – it's mutual replacements or functional substitutions (Harems & Obcowski 2008, NTCIP 2009). That's just what a need of information operations ordering and rational management on the basis of exchanged information systematising and content selection, becomes an urgent and important task.

In many cases such procedures already are executed, nevertheless in numerous ITS applications areas lack of the proper information management and regulations can be observed. It is a result of various reasons, among which lack of necessary or desirable cohesion of the ITS as a whole is one of most important. Elimination of this and other shortcomings requires firstly to identify and systematize information users types (as well human as machines ones) and their needs, then making classification of the types of information, their features, considering even their dimensions and utilitarian meanings. Even defining of the features of technical means necessary or useful for information processing, exchange and presentation is needed. IT is to point that in the last mentioned area, one of main elements influencing system's cohesion and compatibility becomes protocols for inner- and inter-systems communications and interfaces to systems users and surroundings. Such a need can be superbly illustrated by the shortcomings resulted by traffic management systems incompatibility or variety of electronic fee collections along the international routes, from one side, and idea of internationally unified safety supporting eCall system – in other.

2 INFORMATION IN TRANSPORT STRUCTURE

Intensifying and improving quality of the transport related information requires – from technical side – creation and installation of various more advanced devices and programs for information gathering, distribution, processing and usage for inner systems needs and proper improving interoperability between particular systems. Interoperability – first of all – means inclusion by common communication rules the information provision for all of users and operators of all transport systems, enabling distribution of actual, useful and reliable information which can be collected from all possible sources and provided for usage by all interested users, possibly suitably to theirs expectations. It causes a need to pay special attention to information content flowing in telematic systems and between them, especially ensuring optimal solutions applied for execution of these flowing.

Is to be underlined, that optimisation problems are always mostly related to quality of information content, i.e. it's adequacy to time and place of origin, validity and importance, but not as much to technical features of processing and distribution of information.

From the ITS needs point of view, the systems inner information decides about the state and activity of given kind of transport, but important role plays

even outside generated information, describing circumstances and conditions influencing actions of this kind of transport. Of course, for assuring a proper and effective realisation of the transport tasks, there is also need to reach sets of information describing relatively constant (quasi-static) states and circumstances as well as dynamics of occurring processes (Wydro 2009).

It is obvious, that the total amount of information appearing in the system depends on system's dimension, i.e. on numbers of it's elements and processes in it occurring, theirs distraction and geographic locations, on dynamics of these processes and changes in surrounding, but also on types and tasks of the information systems utilizing this information. Also it is reasonable to take for analysis as an area of reference a road transport, which due to its specificity characterized by complexity of roads network with diversity of classes and conditions, states but even managing entities, bearing intensive traffic with high randomness and dependence on environmental conditions, even an area with richest range and diversity of implemented telematic applications, ensures possibly comprehensive analyse of information management problem.

Also, it have to be remembered, that as a result still emerging new technological possibilities, beside new user needs stimulating constructors invention, variety of new telematic applications still is rising, and existing ones use to be essentially upgraded – what together strongly increases demand for information amount and it's improved quality (Report 2009, Wydro 2003). Obviously it broadens also areas of above-mentioned analysis.

With information management questions are also related problems of information transmission (understandable as carrying in space and/or time). What's important, in more and more transport cases, the information have to be delivered to moving objects. Besides, for the sake of required level of the reliability and resistance to possible interferences, some protections means are to be applied, what naturally expands the volume of transmitted information. Such a circumstances brings some difficulties for creation of the information systems, but have to be considered at information categorisation (i.e. problem of confidentiality).

In fact, for various modes of transport can – or may – be applied specialised teleinformatic systems, but, as a rule it's basic structures remains similar, what have some reflection in ITS architectures. Also particular basic applications for information exchange and processing may be equal, what in turn arise legitimacy and need of technical standardisation activity in transport telematics domain. But these last said so far concerns the forms of information, not interfering their contents[1]. If yet the devices should be active with reference to information's content or essence, functioning of such a devices should even be embraced by some defined rules and principles. Also, from infologic point of view, in electronic communications area the kind of transmission technical means is not important, although choice among accessible kinds may have some meaning for reliability, transmission capabilities and costs. Important is however so that information was transmitted in agreed formats (patterns) ensuring mutual understandable communication of system's elements. Having in mind that in telematic solutions becomes needs of communication among:

- Vehicles and infrastructure's teleinformatic equipment,
- Various vehicles,
- Vehicles and informatic and service points or centres,
- Infrastructure's teleinformatic equipment and service points or centres,
- Drivers and related informatic surrounding,
- Informatically co-operating parts of particular vehicles,

may be expected, that will be continued works on integration not only means of information exchange, but even on the manners of these exchange in ITS as a whole and firstly – on information transmission content-oriented protocols and selection and distribution of information methods with striving to more and more necessary automatic languages translations, as well as building personally tailored and dedicated information packages (Gut & Wydro 2010).

2.1 Information sources

In each of information-operated system can be distinguished two main areas of information origin. These are the observed objects and processes delivering basic information and sources of various supporting, already processed information. In transport system as such can be pointed the informational equipment of the transport infrastructure, transport means and entities (persons and institutions) participating in these processes. As examples of infrastructure's equipment delivering primary basic information may be mentioned vehicles detectors or other measuring devices (as photo-radars or weights), weather stations and other environmental sensors, observation systems (cameras), pedestrians detectors, security systems elements and alike. In turn, vehicle's information generating equipment embraces elements of such systems as warning, positioning, emergency (i.e. eCall), movement registration or

even specialised measuring equipment (Floating Car Data). It is worth to underline that contemporary cars use to be equipped with various driver-supporting solutions, as ABS (Anti-lock Braking System), ACC (Adaptive Cruise Control), EBS (Electronic Brake Assist System), ESC, LDWS (Lane Departure Warning Systems), WLDW (Wireless Local Danger Warning) and others (2). These systems actually undergoes to operational integration and delivers information partially used at the time internally in the vehicle, partially transmitted for the outside use, both, in extend appropriate to needs, registered for future use. Next, information delivered by entities participating in transport processes are these generated by persons – individual, corporative or institutional – moving or causing movements of some transport objects.

As mentioned, centres for gathering and processing of raw temporary information, which later is supporting various users of information, form another important group of information sources. By the information processed here is understood operational information used in currently realised transport processes, as well as analytic or reporting ones as for example results of short-, middle- ors long-time analysis. These can be i.e. data from control centres, databases, or managing entities. A good example are sets of information passed to infrastructure's roadside equipment i.e. concerning or applied to traffic control elements as traffic lights or variable message signs, radio announcements and other actual communiqués. Similarly is with information for travellers.

Next, information creating strategies of traffic control in various areas (town, village, roads between inhabited areas) and current circumstances, methods of reaction to particular types of incidents, fleet management and alike, may be numbered among information coming from middle-time analysis. To this class can be included also information collected from observations and registrations of the vehicle's pictures with register plates recognition or points of truck weighing. Hoverer for example prognosis of the traffic flows spread stands for long-time analysis. Distinction between duration of the validity of forecasts important for determination of the sampling frequency of observed processes and observation of it's information content irregularity, seems to be important for information classification patterns.

2.2 Information recipients

Essentially, set of types of information recipients and users is the most meaningful classification criterion for transport information ordering purpose, as types of recipients determines what kind (in meaning of content) and of which quality information is to him needed and when and where have to be deliv-

---

[1] Regulations related to the form of information concerning technical parameters have been known long ago as a "standards" and are properly advanced.

ered. Among information users can be distinguished following main categories (Wydro 2009):
- Rescue services and systems,
- Information and communication systems,
- Administrative institutions,
- Drivers and travellers,
- Corporate operators,
- Research and educational institutions,
- Financial institutions,
- Legal institutions.

Theirs needs decides about basic content structures of used information and schedules as well as conditions of information delivering.

3 INFORMATION STANDARDISATION PREMISES

In the last years in ITS development frames emerged few projects comprising some elements of information content ordering and standardisation. As examples can be mentioned:
- Conception of the Minimal Set of Data (MSD) in eCall system,
- National Transportation Communications for ITS Protocol (NTCIP) project,
- Transport Protocol Expert Group (TPEG) project,
- Open Communication Interface for Road Traffic Control Systems (OCIT) project.

Minimal Set of Data (MSD) brings information necessary to inform rescue services about place, time, circumstances and nature of occurred incident or accident. This information is passed automatically or manually through emergency number 112 to nearest so called Public Safety Access Point (Gut, Wydro 2010) initiating rescue action.

The NTCIP (NTCIP 2009) is a name of American group of standards for communications in transport, specifying open, based on the project participants' agreement, suitable for this communications profiles and protocols as well as common data definitions. These standards allow fulfil all the conditions resulting from needs of communications in the areas of traffic control and transport managements centres.

However TPEG Forum (TPEG), is an European organisation of the group of experts in information technologies, aiming elaboration of the methods and techniques of the collection and delivering for various users – by the broadcasting means (radio, Internet) – information for traffic control and travellers. Here is assumed forming of hierarchically structured information, which recipient will get and will be able to use in various technical means of information processing and also – language independently – by humans. It has to be also information useful for multi-modal transport systems.

Other important accepted assumption is that in information systems structures are not foreseen necessity of building big auxiliary databases, especially in

users receiving devices. Forum tends to develop modular set of tools in prospect standardised by ISO and CEN, taking into consideration possibility of contemporary or future use for various informatic applications.

In traffic management systems particularly important for data exchange organisation are communications protocols. This exchange, usually essential for co-operation of devices and systems provided by various producers, often needs extra investments for building appropriate interfaces and software, what brings significant complications for co-ordination of the systems operations, but even for new functionalities and applications implementation in traffic management structures. In such a cases, as generally in various others information systems, applies a rule of application of "open" protocols. It means application of the protocols worked out and standardised so, that system could work with any device independently of it producer and possess feature of "scalability".

Such a solution presents OCIT protocol (Haremza & Obcowski 2008) being a German standard, but in last years applied in other European countries as an open interface for communication between traffic control systems. OCIT standards are defined for two applications groups. First one, called OCIT-Outstation, pertain communication between local equipment (i.e. traffic lights controllers, measuring stations, VMS) and managing centres. Second one, OCIT-Instation, concerns exchange of information between various applications and systems on the central level of control or management.

Of course an important role in ordering of information areas plays standardisation institutions, mainly international ones like CEN and ISO (ISO).

In both of them activities in ITS (telematic systems) areas are performed by special Technical Committees (TC), each of which is divided between Working Groups (WG), in both cases thematically almost similarly structured. On the basis of commonly accessible information concerning structures of the can be supposed, that in each of Group can be found some elements connected with transport information treatment rules.

4 STANDARDISATION RULES OUTLINE

IT is obvious that basic group of features characterising information used in each well working system are these which can be recognized as determining their utility. Usually assumes (Wydro 2008) that each information in the system have to be:
- Essentially and operationally adjusted to recipient's needs,
- Possibly exhaustive as it concern meaning, completeness and conciseness,
- Ascribed to the time and place,

– Articulated, and in case of transport – easy to language translation,
– Possibly most up-to-dated,
– Verifiable.

As it mentioned earlier, ordering in information management area have first of all to be captured in some classifications frames, what make possible better identification and more convenient operations with their elements. Below are presented fundamental premises for the formation of the frames for content standardisation and information ordering for their more efficient management and usage.

Generally can be accepted, that division of the information features into groups mostly distinctive from the infologic point of view is a proper approach (Wydro 2008). These are features:
– Phenomenological, i.e. universal in relation to any area of application or analysis,
– Social and economic, related to utility in economic or social activity,
– Operational, significant from the point of view of information managing operator or information user.

A an example of phenomenological classification may be quoted a division of information according the following criterions:
– Type of source: inner – external, primary – derivative, public – private
– Kind: quantitative – qualitative, formal – not formal,
– Time: former – actual – future,
– Frequency of occurring: continuous – periodical – incidental,
– Usage: planning – control – decision-making – concluding,
– Level of usage: strategic – tactical – operational,
– Detail level: detailed – summarized – general,
– Presentation form: written – oral – visual.

As economic and social features may be mentioned: a direct market value of information, utilitarian value for economy, accessibility, utility for social activities in various dimensions – cultural, military etc.

From our research matter point of view, the most important is the set of operational features, however others can be also discussed. Analysis of the research matter shows, that legitimated is proposal of classification in two dimensions:
– Areas of applications (utilising),
– Conditions of usage.

Acceptance of the area of applications as the basic classification criterion results from the primacy of meaning and role of information in transport (similarly in any case as in each other branch). For ITS such areas are to be determined by the character of services provided by given system for which given information is necessary. Groups of systems with similar service tasks makes up separated areas of applications. It is to point, that on systems qualifications in some extend influences also technical solutions applied in particular cases, which are often unique from the construction point of view, but shows some universality as can be used in various systems for various goals (i.e. vision systems use to be applied for security levering, traffic control or vehicle recognition). As the systems usually are not mono-functional, ascription them to areas of applications are even not unambiguous. Similarly not unambiguous are qualifications of the areas of applications. These are also qualifications and ascriptions of arbitrary types, even changing with the time. Nevertheless currently these qualifications are quit stable, what seems to be i.e. reflected in the names of Working Groups in relevant Committees of standardisation institutions or research works and papers concerning ITS, as well as in used commonly terminology in professional communication.

What concerns of the formal usage conditions, it easy to state that can be distinguished three categories of obtained or distributed information:
– Obligatory,
– Contacted,
– Free.

It combines with legal rights to information and it's availability, but also with formal conditions related to technical means for information collection, distribution and presentation (Gut, & Wydo 2010).

Undoubtedly it is a factor essential for information operation and requires to be considered in assumption of rules and standards of information operation processes. For completeness of standardisations needs, it is also necessary to give for information (communiqués) some ordered structural form.

4.1 Application areas

Among already numerous telematic systems may be distinguished (Report 2009) basic ones, designed for the provision of single service or fulfilling some particular function (when it work in broader system) and complex ones (integrated) for servicing more complex transport processes on i.e. separated geographical area, mode of transport or tasks group.

Systems of the basic type are numerous and supported on various technical solutions. A good illustration to variety of such a systems, classified on the basis of users needs and contemporary technological possibilities gives list of real service systems presented – among others – in (Wydro 2006), where additionally the systems were grouped with respect to applications areas, though it have to be pointed that the list is not closed as with the time emerges new solutions resulted by new technological possibilities, constructor's invention and users expectations.

Next, as the examples of complex systems can be pointed sets or sub-sets of the basic systems, completed for realization of the complementary functions for fulfilment of the tasks for which they was build. Such a systems are usually ascribed to some given functionality (servicing) areas (IST-FRAME 2004).

According to said above, in particular complex system with well-defined tasks may be distinguished specialised parts, being components of the system as a whole. It is to underline, that specialised systems can in many cases fulfil some additional functions, for example deliver information to other systems.

In proposed standardisation concept assumes that formal classification of the information should be related to concrete telematic systems, with strong consideration of their role in the system and co-operation in functionality area frames, but also with consideration of it's capability to co-operation with other systems, with simultaneous preservation of the development openness and scalability.

Obviously, real classification of the systems from the point of view of information standardisation needs much more deeper analysis.

4.2 *Structural requirements*

As it was mentioned earlier, there is a need to give to information communiqués defined structural form. It is particularly important when are exchanged information between technical devices and even – in some cases – in transmission of the communiqués which have to be of high completeness and precision, as for example it is in the eCall system. Structurally ordered information makes also all the operations concerned with information storage in databases, processing, surveying and analysing. Even transmission of information in agreed formats inside each of systems ensures unambiguous mutual articulation between its elements as well as is necessary for compatibility of different systems. It's the reason for tendency to operations on the ordered sets of dialogs and communiqués and ordered sets and allowed ranges of data, thanks for what not only information users but even telematic system's constructors could communicate in mutually comprehensible and unambiguous manner. It is also obvious need to complete communiqués and other information mails with data pointing place and time of the incident described in this information, and – if it concerns some process - also defining a proper frequency of taking of samples describing a following states of this process (Wydro 2008).

Foregoing remarks allows to state, that in fact the description of the structural form of information means creation of the corresponding meta-information i.e. information about information, which supports, among others, convenience of the identification, absorption and usage of information (Wydro 2008). It suggests elaboration of the system of markers, each of which could be ascribed to par-

ticular category of information and which interpretation would be stored in some database. It could create a convenient in operations, shortened form of above-mentioned description.

4.3 *ACCESSIBILITY CONDITIONS*

The second classification dimension having valid operational meaning is the accessibility status. As mentioned earlier, may be distinguished information, which has to be provided obligatory and cost of which bears operators or administrators of the infrastructure, who bears also responsibility, concerned with regularity of these information and correct delivering. Such obligatory information is for example road signs content, among them – these modern like VMS – or broadcasted by radio or Internet official information. Such ones have to be properly formatted and pass a proper verification procedure, as usage of it may result in material or legal consequences of high significance.

Another category makes information exchanged between partners contracting services containing information as content of the service itself or as a factor influencing essence of the service. Exchange is fulfilled on the basis of the *contract* (agreement) between provider and recipient. As an example may be pointed delivering of roads condition pictures or parking accessibility, performed on the aid of infrastructure administration by some external professional entities. Here also the ranges and formats of information are established, and some legal aspects concerns nor the contents of delivery, but rather assurance of keeping on agreed frequency and continuity of delivery.

At last, there exists also a huge area of free information exchange and provision. As example can be mentioned positioning data (non-professional) delivered by the satellite systems or information broadcasted by CB-radio (even other radios or Internet). In such a cases there is in fact no any formal constrains, and if a recipient undertakes soma decisions or actions based on those information, does it on the own responsibility.

5 CONCLUSIONS

Elaboration of the rules (standards) allowing to order activities in area of obtaining, exchange and usage of the context valuable information should create an important circumstance facilitating functioning, but even construction of the ITS solutions. Such a conclusion comes from survey of contemporary implemented telematic systems as well as from direct discussions in involved professional environment – technicians, researchers as constructors. In many areas of information users such a ordering are in scope of interest of administrations in sense of development and modernizing activities in transport. Therefore elaboration and putting to practice broadly accepted methods of coherent manners of information exchange in transport branch as whole and in ITS particularly, is an urgent question. It should lead to formation of the rational system of operation on content-selected ordered information in transport area.

An important part of above defined task requiring to be researched broadly is a problem of transport meta-information creation and manners of information verification, especially these of high importance for the systems. Other important tasks are reduction of redundancy existing in information by the nature and also caused by information replication, and elimination of unimportant information. Possible solution in these last tasks needs of advanced research with methods of *semantic selection* (Wydro 2008).

Fulfilling of the pointed expectations may be done by adequate research and development entities working in proper interdisciplinary structures and co-operating with international ones. Achieved results in a broader depiction could also make a contribution to methodology of electronic communications systematising and rationalising in other branches of economy, what can be exploited at construction of various development plans in broadly understood electronic communication in information society.

REFERENCES

CEN Technical Committee 278, www.cen.eu/cenorm/sectors/ technicalcommitteesworkshops.

Gut H., Wydro K.B. 2010 eCall system as an equipment of the intelligent vehicle, Magazyn Autostrady, 1-2/2010 (in Polish).

Haremza P., Obcowski D. 2008. OCIT protocol as an international standard enabling integration of the solutions and sub-systems in road traffic control, First Polish Congress of ITS, Warszawa, (in Polish).

ISO Technical Committee 204, www.iso.org/iso/ standards development /technical _committees.

IST-FRAME. 2004. *Planning a modern transport system. A guide to Intelligent Transport System Architecture*, DG T&E, Brussels.

NTCIP 9001. 2009. *National Transportation Communications for ITS Proto*col, American Association of State Highway and Transportation Office, Washington DC.

Report. 2009 *Problems of transport telematics development*, NIT, Warszawa (in Polish).

TPEG, www.ebu.ch/en/technical/projects/b_tpeg.php

Wydro K.B. et al. 2003: *Analisis of the flows of information in ITS* , NIT, Warszawa (in Polish).

Wydro K.B. 2009. *Some problems of information exchange in transport telematic systems*, IX International Conference "Transport Systems Telematics", Katowice – Ustroń.

Wydro K.B. 2006. *Telematic techniques in transport* , Multi-annual program"Development of telecommunications and posts in information socjeit era, NIT, Warszawa (in Polish).

Wydro K.B. 2009. *Inteligent transport systems – an outline of problems*, Magazyn Autostrady, 3, 4, 5/2009 (in Polish)

Wydro K.B. 2008. *Information – features, basic technologies, usage*, Obserwacje - Wydawnictwo Naukowe, Warszawa (in Polish).

8. Application of Fractional Calculus in Identification of the Measuring System

M. Luft, E. Szychta, R. Cioć & D. Pietruszczak
Faculty of Transport and Electrical Engineering, Kazimierz Pułaski Technical University of Radom, Radom, Poland

ABSTRACT: The paper outlines an example of modelling the measurement transducer and actual measurement system with the use of fractional calculus. The algorithm determining these models is presented in the form of a fractional calculus notation and then the models are compared to the ones described by means of classical differential equations. Tests are executed in the programming environment Matlab-Simulink.

1 INTRODUCTION

Modelling of an actual measurement system consisting of many devices requires considering the response dynamics of each of them (Luft & Cioć 2005). When one knows the input signal and the response signal, it is possible to obtain a description of the system dynamics in the form of a differential equation. The accuracy of a thus obtained model depends mainly on the applied method of identification. Application of fractional order derivative-integral calculus (in brief: fractional calculus) for identification purposes provides new possibilities of obtaining a model which reflects the dynamics of the examined object in a more accurate way. (Kaczorek 2009), (Podlubny 1999)

Fractional calculus is not a new concept. It dates back to the 17th century. References to it can be found in the letters of G. W. Leibnitz to de l'Hopital (1695), in the works of L. Euler (1738) or P. S. Laplace (1812). (Ostalczyk 2008)

Many physical phenomena, such as liquid permeation through porous substances, load transfer through an actual insulator, or heat transfer through a heat barrier are described more accurately by means of derivative-integral equations. The dynamics of physical processes such as acceleration, displacement, liquid flow, electric current power or magnetic field flux are modelled by means of differential equations. The courses of these processes are actually continuous variables, $m+1$-fold differentiable, where m is determined subject to the order of the examined fractional derivative. Mass cannot be relocated from one place to another in an infinitely short time. Neither is it possible to change temperature or pressure in an actual object infinitely fast.

A classical notation of the measurement transducer dynamics is based on differential equations, which constitute their mathematical model in the time domain.

Such an equation assumes the following form:

$$A_i \frac{d^{(i)} y}{dt^{(i)}} + A_{m-1} \frac{d^{(i-1)} y}{dt^{(i-1)}} + ... + A_0 y(t) =$$

$$= B_j \frac{d^{(j)} f(x)}{dt^{(j-1)}} + B_{m-1} \frac{d^{(j-1)} f(x)}{dt^{(j-1)}} + ... + B_0 f(x) \qquad (1)$$

where: $y = f(x)$.

In the case of fractional calculus, operators of the function differentiation and integration are combined into one operator D^n.

For differentiation, the n order assumes positive values of $n = 1, 2, 3, ...$ and for integration – negative values $-n = -1, -2, -3, ...$. The neutral operator for the order of $n = 0$ is also defined:

$$D^n f(t) = \begin{cases} \dfrac{d^n f(t)}{dt^n} & \text{for } n > 0 \\ f(t) & \text{for } n = 0 \\ \int_{t_0}^{t}\left[\int_{t_0}^{\tau_1} ... \left[\int_{t_0}^{\tau_{-n-1}} f(\tau_{-n})d\tau_{-n}\right]...d\tau_2\right]d\tau_1 & \text{for } n < 0 \end{cases} \qquad (2)$$

In the classical engineering approach it is only assumed that $t_0 \geq 0$.

In fractional order derivative-integral calculus the arbitrary order derivative is treated as an interpolation of the sequence of discrete order operators (2) by continuous order operators. A notation introduced by H. D. Davis is applied here where the fractional order derivative of the $f(t)$ function can be presented in the following way:

$$t_0 D_t^{\gamma} f(t) = t_0 I_t^{-\gamma} f(t) \qquad (3)$$

where:

t_0 and t - terminals of fractional differentiation or integration;

V - order of the derivative of the integral.

It must be emphasized that in formula (3) the $[t_0, t]$ differentiation range is defined, identical with that which appears in classical definite integrals. Hence, a derivative and an integral of the fractional order are defined in the $[t_0, t]$ range, which in the case of the derivative narrows down to the $[t, t]$ point (range) for the integer order n.

To distinguish between integer and fractional orders, fractional orders are labelled by Greek letters v and μ. For integer orders the commonly applied letters m and n are reserved.

The order of a derivative or an integral satisfies the condition:

$$v \in \mathbf{G}_+, n \in \mathbf{Z}_+ \qquad (4)$$

where: $\mathbf{G}_+$ - set of real, fractional, positive numbers; $\mathbf{Z}_+$ - set of non-negative integers.

In order to emphasize the difference between a derivative and an integral, in H.D. Davis's notation it is written down that $t_0 D^{-v} f(t)$, where the integration order fulfills condition (4), and the minus sign next to the D operator informs that it is the integration operation.

The recent dynamic development of research into the use of fractional calculus for the analysis of dynamic systems encouraged the authors to make an attempt at using it for modelling of both measurement transducers and actual measurement systems. (Luft, Szychta, Cioć & Pietruszczak 2010)

2 IDENTIFICATION ALGORITHM OF MEASUREMENT SYSTEM USING FRACTIONAL CALCULUS

In our work a laboratory system for examining the characteristics of dynamic accelerometers as depicted in Figure 1 was used. The system consisting of an accelerometer, conditioner and a measuring card was modelled. An equation describing the system dynamics was determined by means of the self-regression method with an external ARX input (Cioć 2007). The voltage signal from the end of the examined measurement chain is an identified signal, whereas the comparative signal is the one from the model accelerometer being a response to the sinusoidal input of a generator having the frequency of 200 Hz.

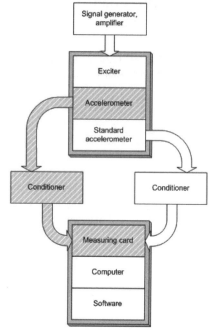

Figure.1. Block diagram of the laboratory measurement system examining measurement transducers.

As a result of the application of the ARX identification method transmittance describing the system dynamics was obtained:

$$G(s) = \frac{0.03215s^2 + 1319.6s + 1.338 \cdot 10^6}{s^2 + 4.678 \cdot 10^4 s + 2.309 \cdot 10^7} \qquad (5)$$

Discrete transmittance (6) for the sampling time $T_p = 10^{-5}s$ was determined on the basis of continuous transmittance (5).

$$G(z) = \frac{0.03215z^2 - 0.05368z + 0.02163}{z^2 - 1.625z + 0.6264} \qquad (6)$$

Transmittance (6) can be written down in the form of a general differential equation:

$$a_2 w_k + a_1 w_{k-1} + a_0 w_{k-2} =$$
$$= b_2 x_k + b_1 x_{k-1} + a_0 x_{k-2} \qquad (7)$$

or as a matrix equation:

$$\begin{bmatrix} a_2 & a_1 & a_0 \end{bmatrix} \begin{bmatrix} w_k \\ w_{k-1} \\ w_{k-2} \end{bmatrix} =$$

$$= \begin{bmatrix} b_2 & b_1 & b_0 \end{bmatrix} \begin{bmatrix} x_k \\ x_{k-1} \\ x_{k-2} \end{bmatrix} \qquad (8)$$

Differential equation (7) when recorded in the derivative-integral form looks as follows:

$$A_2 \Delta_k^{(2)} w_k + A_1 \Delta_{k-1}^{(1)} + A_0 w_{k-2} =$$
$$= B_2 \Delta_k^{(2)} w_k + B_1 \Delta_k^{(1)} x_{k-1} + B_0 w_{k-2}$$

(9)

where $\Delta_k^{(n)}$ is a backward difference of the discrete function defined as:

$$\Delta_k^{(n)} f_{(k)} = \sum_{j=0}^{k} a_j^{(n)} f_{(k-j)}$$

(10)

Having considered expression (10), equation (9) presented as a matrix takes the following form:

$$\begin{bmatrix} a_2 & -a_1 - 2a_0 & a_2 + a_1 + a_0 \end{bmatrix} \begin{bmatrix} \Delta_k^{(2)} w_k \\ \Delta_k^{(1)} w_k \\ \Delta_k^{(0)} w_k \end{bmatrix} =$$

$$= \begin{bmatrix} b_0 & -b_1 - 2b_0 & b_2 + b_1 + b_0 \end{bmatrix} \begin{bmatrix} \Delta_k^{(2)} x_k \\ \Delta_k^{(1)} x_k \\ \Delta_k^{(0)} x_k \end{bmatrix}$$

(11)

Eventually, equation coefficients (9) equal:

$$A_0 = 0.0001$$
$$A_1 = -0.0106$$
$$A_2 = 0.0322$$
$$B_0 = 0.0018$$
$$B_1 = -0.3755$$
$$B_2 = 1$$

(12)

A discrete model determined by the derivative-integral notation takes the following form:

$$G(z) = \frac{z^2 - 0.3755z + 0.001844}{0.03215z^2 - 0.01062z + 0.0001068}$$

(13)

3 COMPARISON OF THE MEASUREMENT SYSTEM MODEL IN DISCRETE TRANSMITTANCE AND DERIVATIVE-INTEGRAL NOTATIONS

The model of the actual measurement system in the form of discrete transmittance (6) was compared to the model in the form of a derivative-integral notation (9) having coefficients (12). Both models were determined on the basis of a continuous model (5) obtained by means of the ARX identification method.

Figure 2 depicts a block diagram of the measurement simulation executed in the Matlab-Simulink package (Matlab®&Simulink®7. 2008. ThesMath-Works™). The simulation was carried out with the use of the ode3 integration method and the sampling time of $10^{-5} s$ for the sinusoidal input signal having the frequency of 200 Hz.

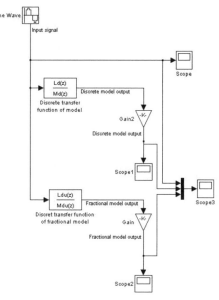

Figure.2. Block diagram of the measurement system simulation.

An accelerometer having sensitivity of 10.18 mV/ms^{-2} was used in the examined system. The model signal was obtained from the reference accelerometer having sensitivity of 317mV/ms^{-2}.

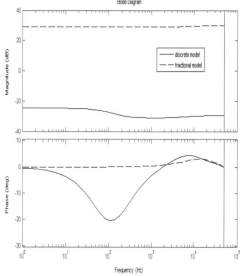

Figure.3. Bode plots for the discrete transmittance model and derivative-integration model.

High amplification visible in the amplitude characteristics of the discrete model (Figure 3) results from the application of voltage signals in this examination. The reference amplification level determined as a ratio of the measurement transducer sensitivity to the model one in the logarithmic scale is $W_r = -29.87 \, dB$. For this value the ratio of acceleration determined on the basis of the measuring sensor voltage signal to acceleration determined by means of a model sensor equals 1. For the determined discrete model a significant phase shift dependent on the input signal frequency can be observed. In the case when the derivative of the integral model was applied, constancy of the phase shift was achieved.

The obtained characteristics (Figure 4) confirm the results obtained from the examination of the amplitude-phase characteristics of derivative-integration model: a reduced phase shift and signal amplitude lower than the model signal. Attention must be drawn to the behavior of the derivative-integration model at the beginning of the simulation. Unlike in the case of the discrete model, the response of this model does not reveal a transitory state resulting from the startup.

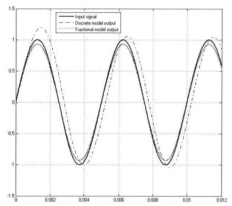

Figure.4. Comparison of the response of the discrete model and the derivative-integration model with an input signal.

4 COMPARISON OF THE CLASSICAL AND DERIVATIVE-INTEGRATION MODELS OF THE MEASUREMENT TRANSDUCER

Analogically, in our examinations we compared responses of the measurement transducer to the sinusoidal input signal.

In the classical notation the dynamics of the examined measurement transducer is written down in the form of the second order differential equation:

$$\ddot{w}(t) + 2\zeta\omega_0\dot{w}(t) + \omega_0^2 w(t) = -\ddot{x}(t) \tag{14}$$

where: ω_0 – natural pulsation, ζ – damping

It is described by three models:
- A continuous model described by the operator transmittance:

$$G(s) = \frac{-s^2}{s^2 + 51s + 255} \tag{15}$$

- A discrete model obtained from a continuous model (15) described by discrete transmittance:

$$G(z) = \frac{-z^2 + 2z - 1}{z^2 - 1.975z + 0.9748} \tag{16}$$

- A discrete model determined by the derivative-integral notation takes the following form:

$$G(z) = \frac{-z^2 + 0.02524z - 6.294e^{-0.005}}{z^2 - 3.161e^{-0.005}z + 1.11e^{-0.016}} \tag{17}$$

Figure 5 depicts a block diagram of the measurement system.

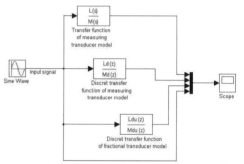

Figure 5. Block diagram of the measurement system for the measurement transducer.

Figure 6 depicts responses of all models of the measurement transducer to the sinusoidal input signal having the frequency of 100 rad/s.

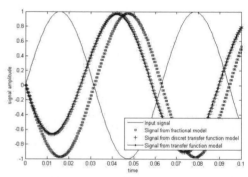

Figure 6. Comparison of responses of the measurement transducer models.

Following the comparison of model responses to the sinusoidal input signal (Figure 6) it can be concluded that: the derivative-integration model of the measurement transducer (17) from the very beginning of the simulation reproduces the input signal amplitude correctly.

The models determined in the classical way - (15) and (16) – reproduce the input signal amplitude correctly after leaving the transient state – in our examinations - after $0.02s$.

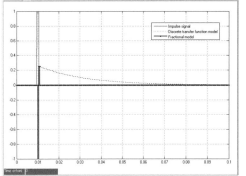

Figure 7. Comparison of responses of the measurement transducer models to the impulse input.

Figure 7 compares responses of the presented models to the impulse input. The classical model response reaches the steady state after 0.08 s from the moment the signal occurs.

In the case of the derivative-integration model this time is reduced to 0.001 s.

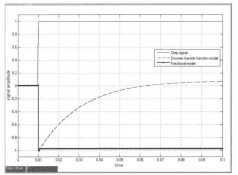

Figure 8. Comparison of the measurement transducer model responses to the step function.

Figure 8 depicts a comparison of the presented model responses to the step function. The classical model response passes into the steady state after 0.06 s from the moment the signal occurs. In the case of the derivative-integration model the steady state occurs after 0.005 s and assumes the value approximating that of the input function amplitude with an opposite sign.

Figure 9 depicts a comparison of Bode frequency plots for the discrete model (16) and discrete model of the measurement transducer determined by the derivative-integral notation (17).

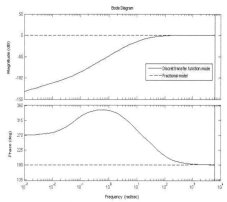

Figure 9. Comparison of Bode plots for models (16) and (17) of the measurement transducer.

The Bode plots (Figure 9) indicate that for the measurement transducer model determined by the derivative-integral method when compared with the model determined in the classical way, the range of the input signal processing is extended by low frequencies. For the presented characteristics, the amplification of the derivative-integration model amplitude equal 0 dB is reached for the frequencies from 0.001 rad/s, and for the "classical" model - from 100 Hz, at a stable phase shift of 180°.
(Luft, Cioć & Pietruszczak 2010. Measurement ...),
(Luft, Cioć & Pietruszczak 2010. Porównanie ...)

5 CONCLUSION

A derivative and an integral of arbitrary order open unimaginable possibilities in the field of dynamic system identification, and creation of new, earlier inaccessible, algorithms of feedback system control. Orders can also be considered as time functions. This also leads to differential equations of variable, time dependent orders.

While applying the derivative-integration model for the creation of the measurement transducer models one obtains models of ideal, in the case of amplitude reproduction, input signal processing.

In the case of the measurement system model, the phase shift obtained was reduced and signal amplitude lower than the model signal.

Moreover, at the beginning of the simulation process, unlike in the case of the discrete model, the derivative-integration model does not have a transient state resulting from the start-up.

It must be emphasized here that the commonly known derivatives are special cases of the calculus presented in this paper. The integer order integral is understood as an equivalent of the definite integral multiplication factor. Therefore we should speak about differential and integral calculus of the integer and fractional order, that is of arbitrary order.

It seems necessary to continue the examinations in order to check how the presented models determined by the derivative-integral method reflect the actual models and whether they reflect the dynamics of the input signal processing more accurately than the models described by the "classical" differential equations.

REFERENCES

Cioć, R. 2007. Korekcja charakterystyk dynamicznych przetworników pomiarowych w diagnostyce wibracyjnej wagonu kolejowego, Doctoral dissertation, Biblioteka Główna Politechniki Radomskiej. Radom, (in Polish)

Cioć, R., Luft, M. 2006. Valuation of software method of increase of accuracy measurement data on example of accelerometer, Advances in Transport Systems Telematics, Monograph, Faculty of Transport, Silesian University of Technology. Katowice.

Cioć, R., Luft, M. 2006. Correction of transducers dynamic characteristics in vibration research of means of transport – part 1 – simulations and laboratory research, 10th International Conference "Computer Systems Aided Science, Industry and Transport", Transcomp 2006, vol.1. Zakopane.

Cioć, R., Luft, M. 2009. Metoda programowej korekcji dynamicznych błędów przetwarzania przetworników pomiarowych, Pomiary Automatyka Komputery w gospodarce i ochronie środowiska, Kwartalnik Naukowo-Techniczny nr 2/2009, str. 22-25, ISSN 1889-6981, Fundacja Nauka dla Przemysłu i Środowiska. Rzeszów, (in Polish)

Jakubiec, J., Roj, J. 2000. Pomiarowe przetwarzanie próbkujące, Wydawnictwo Politechniki Śląskiej.Gliwice, (in Polish).

Kaczorek, T. 2009. Wybrane zagadnienia teorii układów niecałkowitego rzędu, Oficyna Wydawnicza Politechniki Białostockiej. Białystok, (in Polish)

Kaczorek, T., Dzieliński, A., Dąbrowski, W., Łopatka, R. 2006. Podstawy teorii sterowania, WNT. Warszawa, (in Polish)

Luft, M., Cioć, R. 2005. Increase of accuracy of measurement signals reading from analog measuring transducers, Zeszyty Naukowe Politechniki Śląskiej 2005, Transport z. 59, nr kol. 1691. Gliwice .

Luft, M., Cioć, R., Pietruszczak, D. 2010. Porównanie klasycznego i różniczkowo-całkowego modelu przetwornika pomiarowego, Logistyka nr 6/2010, (in Polish)

Luft, M., Cioć, R., Pietruszczak, D. 2010. Measurement transducer modeled by means of classical integral-order differential equation and fractional calculus, Proceedings of the 8th International Conference ELEKTRO 2010, pp TA5_88-TA5_91, (ISBN 978-80-554-0196-6). Zilina, Slovak Republic.

Luft, M., Szychta, E., Cioć, R., Pietruszczak, D. 2010. Measurement transducer modelled by means of fractional calculus, Communications in Computer and Information Science 104, pp.286-295, ISBN 978-3-642-16471-2. Springer-Verlag Berlin Heidelberg.

Matlab®&Simulink®7. 2008. ThesMathWorks™.

Ostalczyk, P. 2008. Epitome of the fractional calculus. Theory and its applications in automatics, Wydawnictwo Politechniki Łódzkiej, ISBN 978-83-7283-245-0. Łódź, (in Polish)

Podlubny, I. 1999. Fractional Differential Equations, Academic Press. New York.

9. Railroad Level Crossing – Technical and Safety Trouble

J. Mikulski & J. Młyńczak

Faculty of Transport, Silesian University of Technology, Katowice, Poland

ABSTRACT: The article brings up the issue of technical and organizational problems connected with railroad crossings. The authors have presented a functional and technical division of railroad crossings. A railroad crossing, or to be precise, a level crossing at grade intersection is a crossing of a railway line by a road where two types of very different vehicles meet. A rail vehicles with high mass on one side, often several hundred times heavier than a motor vehicle, and a road vehicle on the other. Results of a collision are usually predictable.

An increasing number of road vehicles of all types of passenger cars, trucks and truck tractors with semitrailer is accompanied by greater traffic at level crossings and, as a result, a higher likelihood of accidents in their area.

1 INTRODUCTION

Traffic of all vehicles and pedestrians in the area of a crossing of a railway line by a road on one level is based on one basic principle from which all the other ones result – rail vehicles have absolute right of way and each motorist and pedestrian are obliged to respect and exercise particular care, as people and machines sometimes fail.

An increasing number of road vehicles of all types, passenger cars, trucks and truck tractors with semitrailer causes greater traffic congestion at level crossings and, as a result, a higher likelihood of accidents in their area.

Accidents which happen at these particularly dangerous places usually have fatal consequences.

Data published by PKP PLK S.A. [12] concerning accidents at level crossing show that participants in road traffic are responsible for over 90% of accidents due to their failure to exercise proper care while crossing the level crossing or their failure to comply with signals communicated by devices securing the crossing. Therefore, the following questions need to be asked:
- Are level crossings protected in a manner appropriate for the needs?
- What changes should be introduced in level crossing protection systems to increase safety in their area?

These are basic questions commonly asked for some time in most countries on all continents and they equally concern all types of level crossing regardless of their location. In almost every country, making appropriate decisions based on analyses of the actual state of affairs has some influence on legal regulations currently applicable in a given country in the field of protective measures at level crossings.

Generally, legal acts and regulations concerning the construction and maintenance of level crossings take into account over 10 various methods of protecting such crossings and, as a result, warning participants in road traffic approaching to the level crossing. This diversity concerns, among other things:
- Method of controlling warning devices at the crossing, e.g. by means of manual, semi-automatic or automatic control or by automatic detection of the occupied state of an appropriate section of the rail before the crossing caused by an approaching rail vehicle;
- Devices used to protect level crossing, the most important of which included signal indicators, generators of acoustic signals, gates and gate drives equipped with barriers and half barriers;
- The external form (shape and colours) of devices protecting level crossings, e.g. single- double and triple-chamber signal indicators, St. Andrew's crosses of various shapes, sizes and colours, the method of painting the barriers depending on their application in manually-operated and automatic systems, various railway semaphore signals for train drivers etc.

Operational times of devices protecting a level crossing while it is being closed for road traffic must guarantee safety of passage of a road vehicle of the maximum length moving at the minimum speed, if it enters the crossing the moment the warning signal is turned on.

Moreover, the organization of road traffic in the level crossing area is a factor which significantly affects the safety in the level crossing area, in addition to the method of its protection. If a crossroads is situated directly behind a level crossing and the road on which the level crossing is situated is a minor road in respect of the road it crosses, a potential danger occurs consisting in the lack of possibility of leaving the crossing by the participants in road traffic caused by too high traffic intensity on the road with the right of way.

In the situation presented above, proper assessment of the possibility of leaving the crossing safely is possible only after entering the level crossing. A danger may then occur resulting from the fact that road vehicles which have entered the level crossing before turning on the warning system devices will stay there until the time the rail vehicle enters the crossing, which may lead to a fatal accident.

In view of the aforementioned examples of hazards in level crossing areas, a question arises what changes should be introduces to the requirements regulating the rules of protecting level crossings to guarantee the highest possible safety level at crossings and also what requirements in force in individual countries may significantly influence the safety level.

It is usually attempted to increase safety at level crossings by actions conducted simultaneously on various planes. These include:
– Changes in the method of protecting level crossings amounting to changing devices securing level crossings;
– Changes in traffic organization within the level crossing area;
– Changes in the road traffic law and the methods of training future drivers in a way emphasizing problems connected with level crossing safety;
– Enforcement of drivers' compliance with traffic regulations in level crossing areas.
– Changes in the method of funding projects connected with protecting level crossings so that the funding of signalling system devices does not constitute solely the responsibility of railway lines administrator, owing to which the number of projects could be significantly increased;
– Increasing the number of crossings equipped with train detection systems, which, according to the statistics, radically reduces the number of accidents;
– Striving consistently to reduces the number of crossings, especially if such crossings are located near each other.

For the time being, there will probably be no answer to the question whether there will be some form of unification of regulations in this area in EU member states and whether this is possible at all. There are so many differences resulting mostly from various approaches to this issue and also different customs, the mentality of users and the traditions of solving this problem in individual countries.

Also due to an enormous number of level crossings existing in individual countries the importance of the search for optimal solutions in this field is not reduces, on the contrary, it is growing as a result of the natural development of civilization.

2 TYPES OF LEVEL CROSSINGS

In accordance with the applicable regulations 0 level crossings for motor vehicles and pedestrians located within the PKP (Polish Railways) have been divided into the following categories;
– Category "A" – public use crossings with or without gates at which the traffic is controlled by signals transmitted by railway employees;
– Category "B" – public use crossings with an automatic light signalling system and with gates or half barriers;
– Category "C" – public use crossings with an automatic light signalling system without gates or half barriers;
– Category "D" – public use crossings without gates or half barriers and without an automatic light signalling system;
– Category "E" – public use crossings;
– Category "F" – non-public use vehicle and pedestrian crossings

Depending on the requirements resulting from the regulations and traditions specific to a given country, the division into the individual kinds of crossings may vary.

2.1 Optimal density of level crossings

As of 31 December 2008 PKP Polskie Linie Kolejowe S.A. administered railway lines of the total length of 19201 km. This figure includes generally available lines and commercially available ones. There are 16447 public use level crossings and private crossings for pedestrians and vehicles in this network.

The index of the density of PKP level crossings expressed as the average number of crossings per 1 km of railway line in a given country is similar to other European countries.

The optimum density of sites at which linear routes are crossed, including also railway lines, therefore the optimal density of level crossings and viaducts over and under railway lines must be a compromise between financial possibilities, urban planning, social and environmental conditions and safety resulting from both methods of protecting the crossings and the demand for crossing a given railway line.

Table 1. Values of the average number of level crossings per 1 km of railway lines in selected countries

Railway administration	Length of railway line [km]	Number of crossings	Average number of crossings per 1 km of railway line.
Vietnam	2,712	4,842	1.79
Japan	27,230	37,326	1.37
USA	212,400	253,129	1.19
NSB - Norway	4,179	4,805	1.15
the Netherlands	2,808	2,964	1.06
CD – the Czech Republic	9,365	8,684	0.93
PKP - Poland	19,435	14,692	0.76
MAV - Hungary	7,785	5,904	0.76
DB - Germany	36,558	25,941	0.71
SNCB-Belgium	3,471	2,358	0.68
RHK - Finland	5,854	3,521	0.60
France	32,888	19,340	0.59
CFF/SBB – Switzerland	2,975	1,503	0.51
RENFE – Spain	12,310	3,999	0.32
BDZ - Bulgaria	4,320	1,188	0.28
Great Britain	33,800	9,000	0.27
Russia	86,151	13,581	0.16
Iran	6,151	494	0.08

The data presented above prove that the problem of ensuring safety at railway crossings so equipping them with appropriate signalling systems will remain a significant and commonly occurring element of road traffic safety for a long time.

2.2 Modules of a crossing signalling system

Two basic modules can be distinguished in crossing signalling systems:
– a module whose task involves making a decision to turn on/off warning devices ensuring safety in the level crossing area
– a module of implementing appropriate algorithm of crossing signalling devices.

Moreover, crossing signalling systems usually need to be equipped with additional device fulfilling recording, visualizing and remote control functions.

The principles of the operation of the module making decisions about turning on or off the warning system have an immediate influence on the kind of railway crossing. There also exist a large number of staging variants where some devices are manually controlled using special desks while others are controlled automatically.

PKP is considering the possibility of recommending semi-automatic solutions 0 consisting in the fact that in selected category A crossings road signal indicators would be turned on automatically as a result of detecting a rail vehicle moving towards the crossing and gate drives would be controlled by a level crossing attendant (the so-called category A+C crossings).

From the point of view of making a decision about turning on or off the crossing signalling warning system, the following systems can be distinguished:
– automatic crossing signalling systems in which impact device occur in number sufficient for autonomous operation of the signalling system – such systems are usually located on railway routes;
– crossing signalling systems integrated with other rail traffic control systems, such as station systems located near or within railway stations or line block systems in the case of crossings located along the routes.

3 REQUIREMENTS FOR THE WARNING ACTIVATION MODULE IN AUTOMATIC CROSSING SIGNALLING

In general automatic crossing signalling systems can adopt the following, most frequent principles of the operation of the module of turning warning devices on and off:
– train counting principle (PKP, Greece, Croatia, Germany);
– axle counting principle (PKP, Germany, Great Britain, Russia);
– the principle of detecting the head of the train for turning the warning system on and the principle of axle counting or occupied state detection only within the level crossing area (e.g. Greece);
– activation of automatic crossing signalling system by another rail traffic control system (e.g. Poland, Lithuania, Slovakia, the Czech Republic, Croatia) and/or an automatic block signal system (e.g. Croatia);
– manual control concerning all automatic crossing signalling systems or their parts by special desks used by the operating staff.

3.1 Types and configurations of unoccupied state detection

The standard configuration of automatic crossing signalling sensors includes, depending on the number of railway tracks, 1 or 2 turn-off sensors, 1 or 2 turn-on sensors and 1 or 2 departure sensors where in the standard automatic crossing signalling systems it is assumed that the number of tracks at the crossing does not exceed 2 (e.g. PKP) 0.

Turn-off detectors the moment the occupied state is detected should always turn on the warning signalling system regardless of the occupied state of turn-on detectors. After the last train leaves the turn-off sensor and the lapse of additional delay time τ the warning signalling system is turned off.

It is possible to use various types of rail sensors for a given application resulting from the principle of the operation of the signalling system (e.g. wheel sensors can be used as turn-on sensors and a rail short circuit or induction loops on both sides of the road at the crossing can be used as turn-off sensors).

The number of turn-on sensors must be sufficient to determine the occupied status of the individual turn-on zones. Depending on the location of tracks and junctions auxiliary sensors can also be used to determine which track will be the train take at the crossing.

OSE requirements (Greek railways) can constitute an example of the necessity of using several turn-on sensors 0 for hybrid automatic crossing signalling systems combining functions of automatic and manual control implemented by means of a specialized Cp desk.

There also exist turn-on sensors installed close to the crossing whose task involves turning on the signalling in the case of manoeuvres or drives for a replacement signal; in this case the crossing signalling system is turned on earlier for trains which do not stop at the crossing occurs e.g. by means of the signal generated at the station system.

In the general case of requirements:

-n turn-on sensors on one side of the crossing, m turn-on sensors on the other side (the so-called departure sensors) and k turn-off sensors.

The existence of an automatic crossing signalling system encompassing more than 2 tracks in some railway administration may not be excluded. An example here can be requirement concerning selected level crossings at the Kaišiadoris- Radviliškis line (LG Lithuanian railways), where the number of railway tracks per crossing may be more than 2.

3.2 Special purpose sensors

3.2.1 Speed measuring sensor

An example of a special version of a sensor occurring in automatic crossing signalling systems can be an auxiliary sensor for measuring the speed of a rail vehicle over the turn-on sensor placed before the turn-on sensor to create a basic zone with a constant length for measuring the speed of rail vehicle. A wheel sensor is another type of sensor having the ability of measuring the speed.

The speed measurement of a rail vehicle can be used to create a dependency between the preliminary warning time and the speed of the rail vehicle, thus obtaining a crossing signalling system with a quasi-constant warning time. Condition; the rail vehicle must move at a constant speed in the zone between the turn-on measuring sensor and the crossing, in particular, it may not be accelerating as this may result in a dangerous situation.

3.2.2 Sensor on siding track

There exist three-track crossings where the third track fulfils the function of a crossing. 0. In this case the signalling system is activated manually on this track and the warning may be turned off manually or automatically (PKP). In the latter case a turn-off sensor is needed at the crossing in the siding track. The method of turning off the warning system by a vehicle on the siding track is usually determined by the designer of the signalling system.

3.2.3 Sensors establishing category A crossing signalling

In the case of manually controlled crossings (e.g. category A crossings) 0 located on the route near a station, if the distance between the crossing and the station is insufficient to install TOPs over there and the crossing must be released by rail sensors located at the crossing. In this case category A crossing by A drivers should have one sensor on each track whose task is to determine the time when the gate should be turned on after the passage of the last carriage and making it possible to open them by the level crossing attendant.

In the case of a category A crossing located within a station there can be any number of tracks per crossing 0 and the demand of fixing the crossing at a given position as well as releasing it is implemented only by detecting the occupied state of the section by the station system. In this case no sensors are located within the crossing in the system.

3.3 Time of turning on automatic crossing signalling warning

The moment the train leaves the turn-on sensor a special timer can be activated - measuring the so-called automatic crossing signalling deactivation time (e.g. Croatian and Greek railways). If the train does not arrive at the crossing in a specified time the turn-in warning signal coming from this train is erased and the automatic crossing signalling system goes to the waiting time status. It should be noticed that there can be two such timers. The first is responsible for trains moving along track 1 in the basic direction and along track 2 in the opposite direction, while the other one is responsible for trains moving along track 2 in the basic direction and along track 1 in the opposite direction. (e.g. Croatian railways)> In general, the possibility of differentiating these times must be adopted not only as far as the direction is concerned (as above) but also as regards the track and generally the crossing.

There also exist requirements regarding the conditions when the deactivation time is turned off. E.g. in the case of the Croatian railways the deactivation time does not start when the train stops because of block signal or when the train occupies the crossing zone. If the block signal is turned on afterwards, the time of turning the signal on may be of any length until the train leaves the crossing. Generally, the starting and stopping conditions of automatic crossing signalling activation time can be very different for various users.

In the simplest case it is a time interval specified in advance valid for all directions, tracks and crossings (e.g. this time is 210 s for Greek railways).

According to the specification of some users' requirements (LG, OSE), the time of turning the signal on can be limited not only in advance but also in arrears.

In accordance with Lithuanian requirements an entry of a train on a turn-on sensor towards the crossing blocks the turn-off sensor for 40 seconds, which is aimed at ensuring that no accidental turn-off sequence occurs at this sensor before the train passes. It means that the automatic crossing signalling system will be turned on for at least 40 seconds, even if the train passes the crossing sooner. With each subsequent train detected by the turn-on sensor this time interval is counted from the beginning.

In general, there exist requirements as regards the minimum and maximum value of the time the warning of the automatic crossing signalling system is turned on in connection with a train, and turning off the signalling after the maximum turn-off time as well as the lack of the possibility of turning it off earlier before the lapse of the minimum time can involve additional conditions depending on the location of the level crossing.

4 REQUIREMENTS FOR THE WARNING MODULE IN AUTOMATIC CROSSING SIGNALLING

According to the most general requirements 0 (e.g. according to PKP or the German railways), it should be possible to group warning devices so that e.g. 1, 2, 4... devices of the same kind could be declared in each group. Groups of devices are usually subordinated to individual control channels of the automatic crossing signalling system. Feedback signals from the individual groups of devices should be available in both control channels so the information about the entire system is available in the other working control channel even if one controller does not work (e.g. according to German railway requirements).
- The maximum numbers of devices (e.g. according to PKP requirements)) 0, which it should be possible to connect in the automatic crossing signalling system are the following:
- A level crossing across no more than 4 tracks, from which the maximum number of impact devices results – train sensors (12 at most) or wheel sensors (16 at most);
- 20 road signal indications in a maximum of 4 groups;

- 4 acoustic signal indications in a maximum of 2 groups;
- 12 gate drives in a maximum of 6 groups;
- 8 railway semaphore signals controlled independently in 1-4 groups.

In a given group of devices control of common output signals from the controller can be assumed as well as feedback signals separate for each device or common for a given group depending on the ordering party's requirements.

4.1 Road signal indicators of the automatic crossing signalling system

Road signal indicators are turned on whenever the signalling system is turned on. The road signal indicators are usually turned off, when it is possible to enter the crossing by road users due to the absence of rail vehicles at specified crossing zone sections and as a result of the appropriate location (status) of the other warning devices, i.e. when the signalling system is in the waiting mode and gate drives are in their upper position, or after the lapse of the adopted time interval measured from the moment of the activation of the signal setting the gate drives in motion towards the opening of the crossing, after which the gate drives should be in their upper position.

Table 2. Road signal indicators according to railway requirements in various countries **Błąd! Nie można odnaleźć źródła odwołania.**

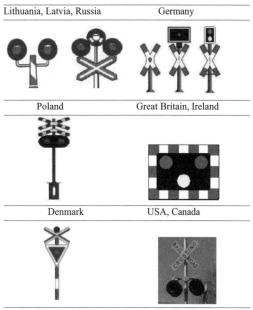

Lithuania, Latvia, Russia	Germany
Poland	Great Britain, Ireland
Denmark	USA, Canada

Among requirements concerning road signal indicators in various countries the following indicators can be distinguished [Table 2]:

- Two-chamber horizontal signal indicators warning with pulsing red light.
- Two-chamber vertical signal indicators warning with constant yellow and red light in a way analogous to traffic lights at crossroads.
- Three-chamber signal indicators warning with pulsing red light or pulsing white light informing the road users about the working order of automatic crossing signalling devices and the unoccupied status of track sections in the area of the crossing.

Road signal indicators are usually equipped with additional warning signs in the form of St. Andrew's crosses whose shape and colours must comply with the requirements applicable in a given country [Table 3].

Table 1. Variant of St. Andrew's crosses used in connection with road signal indicators of automatic crossing signalling systems.

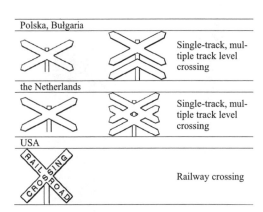

Austria		Multiple track level crossing
Belgia		Single-track, multiple track level crossing
Sweden		Multiple track level crossing
Polska, Bułgaria		Single-track, multiple track level crossing
the Netherlands		Single-track, multiple track level crossing
USA		Railway crossing

The following requirements apply as regards the source of light used in road signal indicators:
- One-filament bulbs – there exists a common requirement amounting to the necessity of detecting a burnt-out bulb and a shorted bulb, both when the automatic crossing signalling system is in the warning and in the waiting mode (e.g. PKP);
- Two-filament bulbs – there exists a common requirement amounting to the necessity of detecting a burnt-out filament and a shorted filament of each road signal indicator, regardless of the signalling system mode, If one of the filaments is damaged (burnt-out or shorted) the system must switch the lighting to a spare filament and to send information about the occurrence of a fault consisting in the lack of a spare filament in a given road signal indicator (e.g. Croatian, Greek, German railways).
- LED – the requirement for working order inspections of the lighting of the indicator chambers amounts to ensuring the possibility of detecting a fault consisting in decreasing light intensity by a defined value in the case of lighting chambers consisting of a large number of LEDs. If light sources consisting of one or several LEDs are used, a working order examination based on the optical feedback is often required.

In various countries various variants of St. Andrew's crosses are used in connection with marking level crossings. These signs can be encountered bout at crossing without any signalling devices and at crossing with various types of signalling crossing devices, also including automatic crossing signalling systems.

4.2 Acoustic signal indicators of automated crossing signalling systems

Bells, buzzers or other acoustic signal generators are turned on together with turning the signalling system on or when another train enters the crossing zone when the signalling system is still turned on and the bell has been already turned off.

The system is turned off according to the following assumptions which are different for individual users;
- when the head of the last train moving at any section between the turn-on sensor and the crossing and headed towards the crossing enters the turn-off sensor area (e.g. PKP);
- when all gate drives are in their lower position (e.g. German and Greek railways).

The bell or another acoustic signal generator generates a signal with constant intensity and frequency.

The inspection of the acoustic signal generator e.g. by checking the presence of current in the executive system of the executive element is often optional (e.g. PKP). (in the future it may be necessary especially at category C crossings 0).

According to the requirements of some railway administrations (e.g. British and German railways) the acoustic signal generator should be able to generate a signal with two different intensity levels and two different frequencies when it is turned on.

4.3 Gate drives of automatic crossing signalling systems

The required algorithm of gate drive operation is often different for individual groups of drives. There can be entry and exit drives.

Gate drives can be locked or unlocked in the lower position.

There is one or two groups of gate drives. Each group usually contains two gate drives but there can also be one or three drives within one of the groups.

After the time of preliminary warning until the signalling system is turned on the entry gate drives are started (they usually close the right halves of the road in countries with right-hand traffic) and the passage of barriers to the lower position. In the case of two groups of gate drives, the other group of exit drives (they usually close the left halves of the roadway) starts from the upper position when all barriers from the first group reach their lower position, but no later than at a specified time which has lapsed from the time the road signal indicators where turned on.

If unlocked drives in their lower position will lose this position (e.g. they are lifted by a road user), the control system should close the gate drives until they reach the lower position again after a specified delay time (e.g. 1 min.).

When the signalling system is turned off, the gate drives immediately move to the upper position.

The required position of all drives (upper and lower) must be achieved in an non-exceedable time, the value of which is determined by the user's requirements.

If the voltage in the gate drives power supply is cut off, they should reach the lower position by free falling (e.g. according to the requirement of PKP and the Lithuanian railways) or it is forbidden (e.g. according to the requirements of Greek railways).

4.4 Barriers of automatic crossing signalling systems

Barriers controlled by gate drives connected to a given control system are inspected for their continuity. The absence of the barrier continuity system informs that they have been broken e.g. by a road user or by strong wind. The absence of the barrier continuity signal is also a fault turning on the forbidding signal at appropriate TOPs in connection with turning on the warning system.

Barriers are usually equipped with lights which emit pulsing and continuous light, when the crossing signalling system is in the warning mode (e.g. PKP, British railways). Barriers can also have reflective belts (e.g. according to the requirements of Greek and Lithuanian railways).

4.5 Obstacle detectors at level crossings

In the case of automatic crossing signalling systems equipped with barriers closing the entire width of the road there exist requirements concerning especially selected level crossings for such systems to have obstacle detectors at level crossings. The task of obstacle detection systems at level crossings is to detect an obstacle (persons, vehicles) at the crossing where such persons or vehicles were closed between the barriers as a result of a coincidence. The signal about the existence of an obstacle may be used in the track-vehicle system to automatically turn on the train braking or to inform the train driver about approaching to an obstacle, also by appropriate indications displayed at semaphore signals.

4.6 Integration of the automatic crossing signalling system with traffic lights

There exist general requirements of interaction between crossing signalling and road signalling systems for individual locations of level crossings as regards crossroads. The requirements of German railways are a good example here 0. If there is a danger that the traffic at the crossing will be obstructed as a result of traffic accumulation because of a nearby crossroads equipped with traffic lights, a dependency between the crossing signalling system and the traffic lights should be introduced.

5 CONCLUSIONS

Each year on average 330 fatalities were recorded as a result of accidents at level crossings in EU countries in the years 1996-2000. At the same time on average 40 000 people died in car accidents each year. Although the number of accidents at level crossings does not exceed 1% of the total number of all road accidents, however, accidents at level crossings have the features of transport catastrophes due to a large number of fatalities, hundreds of severely injured persons and huge financial losses.

An expert evaluation performed by the largest rail traffic operator in the EU – the German railways DB clearly indicated that the reason for numerous accidents happening at level crossings is the ignorance of individual warning signals of the crossing signalling systems and the drivers' lack of the awareness of breaking the law. The marking methods and the crossing signalling systems used in the individual EU countries are significantly different, which makes them incomprehensible for some road users. Colours of the signs, their shape, location and the importance of light and acoustic signalling systems is different in various EU countries.

The necessity of taking actions to decrease the number of accidents at level crossings is usually emphasized and discussed in the media in connection with particularly spectacular catastrophes which occur almost every year.

The report prepared as a result of the aforementioned DB expert evaluation published in December 2003 contains the following recommendations aimed at reducing the number of accidents at EU level crossings:

- Exchange of accident data at level crossings between EU countries to increase the effectiveness of risk analysis and to define remedial measures;
- Exchange of data on changes of parameters of level crossings, including the intensity of rail and road traffic, the location of level crossings relative to roads and crossroads and updating the classification of level crossings in a give category according to the adopted classification;
- Standardization of the level crossing categorization criteria across the EU;
- Registration of the behaviours of participants in road traffic in connection with accidents;
- Selecting level crossings where particularly serious accidents occur and equipping them with light signalling systems and gates, which reduces the risk of accident occurrence by as much as 90%.
- Implementation of appropriate regulations aimed at standardization of the principles of the operation of crossing signalling systems in EU countries;
- Devising and implementing new elements of crossing signalling systems increasing the safety level and eliminating the manual control, such as obstacle detectors at level crossings, ATP systems;
- Undertaking a media campaign informing about actions aimed at increasing the safety level at level crossings which have been taken in the entire EU;
- Spreading information about the catastrophes which occurred, their reasons and preventive measures which have been undertaken;

- In the long term, the application of systems of data processing and transmission on rail vehicles to inform road users about a train approaching to the level crossing and to inform the train driver about the signalling system status at the level crossing.

Safety analyses at crossings in selected developed and developing countries have recently led to the following conclusion - Countries in which the safety level at level crossings is relatively high are countries where the percentage of level crossings equipped with automatic or manually controlled crossing signalling systems is the highest.

It results from the information presented above that warning signalling systems are indispensable for increasing the safety at level crossings.

LITERATURE

[1] Technical guidelines for the construction of traffic control devices in the enterprise PKP WTB E-10 (Ie-4)
[2] Regulation of Ministry of Transport and Maritime Economy of 26 February 1996 on the technical requirements to be met from the crossing of railway and public roads and their location (Dz. U. Nr 33 poz. 144 z dnia 20.03.1996 r.).
[3] Amendment to Regulation (draft) - The Regulation of the Minister of Infrastructure on the technical requirements, which should correspond to the newly built, upgraded and existing crossing of railway lines and roads and their. BAiT 2005.
[4] Automatic level crossing system model SPA-41. Operation and Maintenance Manual. DTR-97/SPA-4/ko. BT ZWUS.
[5] Automatic level crossing system model SPA-5. Operation and Maintenance Manual. DTR-2006/SPA-5. BT ZWUS.
[6] Automatic level crossing systems fort he Hellenic Railways Organization. Purchasing specification. Issue 2. August 2000.
[7] 12Bahnübergangsalagen planen and instandhalten; Technische Sicherungsanlangen an Bahnübergangen anordnen. Ril 815.0032
[8] Funktionsbedingungen der Stellwerkslogik und der in das ESTW itegrierten Techniken; Teilheft F2: Signalgesteuerte Bahnübergangs – Sicherungseinrichtungen (SBÜE).
[9] Automatic level crossing system model SPA-4/EOC-4. Operation and Maintenance Manual. DTR-2002/SPA-44. BT ZWUS.
[10] Trends in the development of level crossings from a Polish point of view. Marian Malcharek / Piotr Cygoń. Signal und Draht. 2004/4.
[11] Crossing Monitoring and Detection System. CrossGuard. GE Transportation Rail. Doc. 30005-A.
[12] Annual Report of Polish Railway Lines (PKP-PLK S.A.) 2008.

10 Application of the Polish Active Geodetic Network for Railway Track Determination

C. Specht & A. Nowak
Polish Naval Academy, Gdynia, Poland

W. Koc & A. Jurkowska
Gdansk University of Technology, Gdansk, Poland

ABSTRACT: One of the main problems related to railroad surveying design, and its maintenance is the necessity to operate in local geodetic reference systems caused by the long rail sections with straight lines and curvatures of the running edge. Due to that reason the geodetic railroad classical surveying method requires to divide entire track for a short measurement sections and that caused additional errors. This problem has been resolved after the evaluation of the active geodetic networks in different countries which used Virtual Reference Station (VRS) idea, General Packet Radio Service (GPRS) transmission method and Networked Transport of RTCM via Internet Protocol (NTRIP) defined in 2004 by The Radio Technical Commission for Maritime Services (RTCM). In 2009, at the Polish Naval Academy and Gdansk University of Technology have been carried out, for the first time, continuous satellite surveying of railway track by the use of the relative phase method based on geodesic active network ASG-EUPOS and NAVGEO service. The article presents surveying results of the measurement campaigns realized and its application for railway track determination. The accuracy analyzes and results of the implementation in railroad design process are discussed.

1 INTRODUCTION

Permanent observations GNSS implemented by large-area satellite network have been transformed into complex data communication system. They have offered not only post processing GPS services but also the provision of the correction data sent at real time. The first stages of their development were national passive systems created at the beginning of the 1990s, also in Poland [Baran L. 1994]. They have evolved from single reference stations located at technical universities to national systems. They were characteristic of autonomous station, lack of standardization within the range of unified report data replacement exploitation and local character of use [Dziewicki M. et al 1998]. As time passed by, passive systems gained differential function (GPS) of real time, becoming active structures - making DGNSS services possible. Meaningful broadening zone action, similar to nautical DGPS system, was connected with new telegram type RTCM [RTCM 2004b].

The first idea of creation permanent station network GNSS was prepared in 1995 on the initiative of Komisja Geodezji Satelitarnej Komitetu Badań Kosmicznych i Satelitarnych PAN and Sekcja Sieci Geodezyjnych Komitetu Geodezji PAN [Baran L., Zielinski J. 1998]. It was assumed that the network should be multifunctional and adapted not only to geodesy. In the consequence of different centres activity, the local stations were forming. They were created in Warszawa, Łódź, Gdańsk and at intensive mining industry area (Górny Śląsk, Lubińsko-Głogowski Okręg Miedziowy) [Baran L. et al 2008]. Then a six-point network at Śląsk and a three-point network at the Three-City area were created [Ciecko A. et al. 2003]. The dominant world trend at the beginning of the 21 century was starting active national network activity, for example CORS, SAPOS, SWEPOS, OS-AGN. The networks offered users payable or unpaid services as well as payable real time services [Cord-Hinrich J. 2008; Rizos C, et al 2003]. Modernity of network techniques, compared to classical coordinates determination with exploitation single reference station and movable receiver in RTK method, lies among other things in implementation of correction using virtual reference station VRS . It enables working out of pseudo-distance correction dedicated to receiver coordinates.

In 2007 year Główny Urząd Geodezji i Kartografii started Active Geodetic Network ASG-EUPOS which is the national permanent GNSS stations network offering services for geodesy and navigation [Bossy, et al 2008]. That investment was ended in April, 2008 with service and data communications tests.

2 ASG-EUPOS

In the 1990s Central and East European countries started setting up DGNSS stations. They were not compatible with the West European stations. In Berlin, in 2002 was taken a decision about European Position Determination System development in the direction of the East. The Polish part of the system ASG-EUPOS has consisted of 98 reference stations evenly located over country(Fig. 1). Except for the new starting up stations, the system has adapted also existing stations managed by universities, research and development centres, state administration and private firms. At the present moment ASG - EUPOS is composed of the next reference station groups: 84 stations with GPS module, 14 stations with GPS/GLONASS module. Additionally the system has cooperated with nearly 30 foreign stations.

Figure 1. ASG-EUPOS architecture.

National Management Centres, called also Counting Centres are the second segment of ASG-EUPOS system. Central Office is in Warszawa and its branches are in Katowice. They have a task of controlling and managing stations network, correction generating of made observations and making satellite surveying available to the recipients. All interferences are signalled and analysed; if necessary countermeasures have been undertaken. Both Counting Centres are redundant in the range of done services. Beside position surveying services Counting Centre has maintained reference system. Weekly control makes possible supervising the invariability of points defined the system. The highest number of users served by Counting Centre is 1200. Users making use of centre services at provision correction data real time exploit mainly Internet and GSM. Corrective information is sent to users via Internet using specially elaborated NTRIP protocol. GSM has used package data transfer GPRS. Working far from the cities can present a problem of being out

within range of one or even all mobile communications operators. Because of that, users who want to take advantage of the provision correction data should have SIM cards of a few operators.

3 MEASUREMENTS

In 2009, the Team have been carried out, for the first time, continuous satellite surveying of railway track by the use of the relative phase method based on geodesic active network ASG-EUPOS and NAVGEO service. Still continuing research works focused on the GNSS multi-receivers platform evaluation for projecting and stock-taking. Next year the same team repeated similar measurements (07.04.2010) on the railroad between two Gdansk stations (Gdansk-Central Station and Gdansk New Port Station). Three Leica GPS Total station system 1200 SmartRover (with ATX1230 GG antenaes) receivers were located in the diameter of the platform (Fig. 2). Polish Active Geodetic Network ASG-EUPOS was used as a reference network transmitted Real Time Kinematic Positioning Service according to RTCM 3.1. standard.

Figure 2. GPS Total station receivers location on the platform.

For the best measurement time period determination software Leica survey design was used. It shoved the best constellation for GPS satellites were between 10.40-11.40 LMT and 12.40-14.430 LMT. The GPS system guaranteed 8-9 satellites with PDOP coefficient close to 1 (Fig. 3).

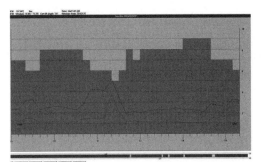

Figure 3. Satellite availability and PDOP values during the measurements.

In order to evaluate accuracy characteristics of different GNSS methods received positions were recorded. Three GNSS receivers used different correction method: VRS, FKP, MAC, in the way Gdansk Central Station to Gdansk-New Port. In the opposite site all receivers used VRS corrections. Previous measurement], have been shoved that 30 cm distance between positions was enough for sample rate. Data files were containing position time series in the format of NMEA 0183 standard, GGA referenced to WGS-84 datum ($a = 6378137.00$ m, $b = 6356752.314$ m).

Measured ellipsoidal coordinates were transformed to Gauss-Kruger (X,Y) conformal coordinates, based on relations:

$$x = R \cdot \left[dL \cos(B) + \frac{dL^3}{6} \cos(B)^3 \left(1 - t^2 + \eta^2\right) + \right.$$
$$\left. + \frac{dL^5}{120} \cos(B)^5 \left(5 - 18t^2 + t^4 + 14\eta^2 - 58\eta^2 t^2\right) \right], \quad (1)$$

$$y = k \cdot R \left[\frac{S(B)}{R} + \frac{dL^2}{2} \sin(B)\cos(B) + \frac{dL^4}{24} \sin(B) \cdot \right.$$
$$\cdot \cos(B)^3 \left(5 - t^2 + 9\eta^2 + 4\eta^4\right) + \frac{dL^6}{720} \cdot \quad (2)$$
$$\left. \cdot \left(\sin(B)\cos(B)^5\right) \cdot \left(61 - 58t^2 + t^4 + 270\eta^2 - 330\eta^2 t^2\right) \right]$$

where:
B, L - measured ellipsoidal coordinates,
R - radius of curvature in the prime vertical,
$S(B)$ distance from the equator to defined coordinate B,
dL - difference in longitude between L and prime meridian,
$k = 0.999923$ – scale factor.

Others parameters could be calculated as:

$$t = \tan(B), \quad \eta = \frac{e^2 \cos(B)^2}{1 - e^2} \quad (3,4)$$

where:
e - eccentricity of ellipsoid.
η - orientation angle of distortion ellipsis.

4 RESULTS

On the basis of the obtained results it was possible to define the main direction of the whole railway route and its segments. These are basic data to design a railway geometric system. For that purpose, a particular run of the route was examined on railway segments. One of the segments – straight line (230 fixes) was used for regression analyses. All calculations were realized in Mathcad, software version 13. Least Squares Method were used for analyses (LSM).

The better estimate than Least Square Method can be obtained using weighted least squares (WLS), also called generalized least squares (GLS). Weighted least squares come from different subpopulations for which an independent estimate of the error variance, if available. The idea is to assign to each observation a weight that reflects the uncertainty of the measurement. In general, the weight p_i assigned to the i th position observation y_i, x_i, will be a function of the variance of this observation. For railway measurements each GPS solution was characterized by 2D and 3D position standard deviation. The linear regression function can be estimated by finding values of a and b minimalizing:

$$p_i \left(x_i - \bar{x}_i\right)^2 = \sum_{i=1}^{n} p_i \left[x_i - (a + by_i)\right]^2 = \min \quad (5)$$

The estimated line coefficients (slope and intersection) a, b can be calculated as follows :

$$\bar{a} = \frac{\sum_{i=1}^{n} p_i x_i \sum_{i=1}^{n} p_i y_i - \sum_{i=1}^{n} p_i \sum_{i=1}^{n} p_i x_i y_i}{\left(\sum_{i=1}^{n} p_i x_i\right)^2 - \sum_{i=1}^{n} p_i \sum_{i=1}^{n} p_i x_i^2}, \quad (6)$$

and

$$\bar{b} = \frac{\sum_{i=1}^{n} p_i x_i \sum_{i=1}^{n} p_i x_i y_i - \sum_{i=1}^{n} p_i y_i \sum_{i=1}^{n} p_i x_i^2}{\left(\sum_{i=1}^{n} p_i x_i\right)^2 - \sum_{i=1}^{n} p_i \sum_{i=1}^{n} p_i x_i^2}. \quad (7)$$

All GNSS recorded track had 3924 m (one way), but for analyses the straight line (200 m) was chosen. The statistical and distribution results for the two trials are presented in Tables 1 and 2.

Table 1. Statistical result of the straight line determination by 3 receivers working in VRS mode (one way)

receiver	receiver no 1		receiver no 2		receiver no 3	
Type of correction	VRS		VRS		VRS	
	Distribution Statistics					
	X1	Y1	X2	Y2	X3	Y3
Mean $[\cdot 10^6\ m]$	6.0263569621	6.5415907669	6.0263571704	6.5415906016	6.0263567424	6.5415909327
Median $[\cdot 10^6\ m]$	6.0263570739	6.5415906803	6.0263572673	6.5415905159	6.0263568331	6.5415908585
Stand. Dev $[m]$	38.7619605809	29.1948618444	38.7493257421	29.1851593437	38.7375729450	29.1756706780
Variance $[\cdot 10^3\ m]$	1.5024895881	852.339958115	1.5015102455	851.773525919	1.5005995577	851.2197595088
	Regression Statistics					
Intercept $[\cdot 10^7]$	1.4711613582		1.4711668835		1.4711858504	
slope	-1.3276979453		-1.3277063933		-1.3277353859	
Correlation coeff	-1		-1		-1	
R^2	0.9999998712		0.9999998404		0.9999998669	
Covariance $[\cdot 10^3]$	-1.1253978011		-1.1246570612		-1.1239504268	
Standard Error	0.0139449256		0.0155134730		0.0141660897	

Table 2. Statistical result of the straight line determination by 3 receivers working in the: FKP, VRS and MAC mode (opposite way)

receiver	receiver no 1		receiver no 2		receiver no 3	
Type of correction	FKP		VRS		MAC	
	Distribution Statistics					
	X4	Y4	X5	Y5	X6	Y6
Mean $[\cdot 10^6\ m]$	6.0263562204	6.5415913185	6.0263562099	6.5415913365	6.0263562099	6.5415913365
Median $[\cdot 10^6\ m]$	6.0263560976	6.5415914045	6.0263561257	6.5415914066	6.0263561257	6.5415914066
Stand. Dev $[m]$	38.8605001350	29.2778413408	38.9118236496	29.3172109044	38.9118236496	29.3172109044
Variance $[\cdot 10^3\ m]$	1.5101384707	857.191993576	1.5141300197	859.498855215	1.5141300197	859.4988552156
	Regression Statistics					
Intercept $[\cdot 10^6]$	1.4711613582		1.4708806684		1.4708806684	
slope	-1.3276979453		-1.3272688598		-1.3272688598	
Correlation coeff	-1		-1		-1	
R^2	0.9999998893		0.9999998683		0.9999998683	
Covariance $[\cdot 10^3]$	-1.1314655745		-1.1344833801		-1.1344833801	
Standard Error	0.0129660370		0.0141600086		0.0141600086	

The lower figure present plot of positions determined by six GNSS receivers on the same 200 meters part of the rail road.

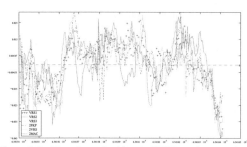

Figure 4. Positions determined by six GNSS, receivers on the same 200 meters part of the railroad.

Application of the Weighted Least Square Method allows railway track line determination with accuracy 1-1.5 cm (fig. 5).

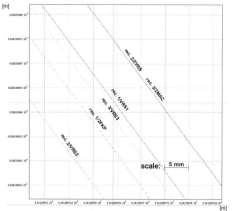

Figure 5. Railway Track determined lines calculated by Weighted Least Square Method.

5 CONCLUSIONS

At a present stage it can be found that implemented ASG-EUPOS in the railway route determination gives researchers a whole new perspective on applied research. Its use enables very precisely basic data definition for design modernization of the railway line.

Calculated distribution and regression statistics for all six receivers shown high accuracy (1-1.5 cm) of the railway route determination based on GNSS active geodetic network measurements.

The main reason of the determining errors position and difference existence between receivers were conditions of geometric observation represented by geometric coefficient DOP.

Resolving tending towards minimal mean error position with ASG-EUPOS usage should be searched in measurement planning stage with maximum number of available satellites and minimal value of DOP coefficient.

REFERENCES

Baran W. 1994. New National System of Geodetic Coordinates in Poland. *Geodezja i Kartografia 1994, t. XLIII, z. 1, p. 41-49.*

Baran L., Zieliński J. B. 1998. Active GPS Stations as a New Generation of the Geodetic Network. *Geodezja i Kartografia 1998, t. XLVII, z. 1-2, p. 33-40.*

Baran L. W., Oszczak S., Zieliński J. B. 2008. Wykorzystanie technik kosmicznych w geodezji i nawigacji w Polsce, *Nauka 2008, nr 4, s 43-63.*

Bosy J., Graszka W., Leonczyk M.. 2008. ASG-EUPOS The Polish Contribution to The EUPOS Project, *Symposium on Global Navigation Satellite Systems*, Berlin, 11-14 November.

Ciećko A., Oszczak B., Oszczak S. 2003. Determination of Accuracy and Coverage of Permanent Reference Station DGPS/RTK in Gdynia. *Proceedings of the 7th Bilateral Geodetic Meeting Italy – Poland.* Bressanone, Italy, Reports on Geodesy, nr 2 (65), 2003 p. 45-51.

Cord-Hinrich J., 2008. SAPOS-Part of a Geosensors Network. *Symposium on Global Navigation Satellite Systems.* Berlin, 11-14 November

Dziewicki M., Felski A., Specht C.. 1998. Availability of DGPS Reference Station Signals on South Baltic. *Proceedings of the 2nd European Symposium on Global Navigation Satellite Systems – GNSS'98.* Toulouse, France.

Rizos C., Yan T., Omar S., Musa T., Kinlyside D. 2003. Implementing network-RTK: the SydNET CORS infrastructure. *The 6th International Symposium on Satellite Navigation Technology Including, Mobile Positioning & Location Services.* Melbourne SatNav 2003, Australia.

11. The Advantage of Activating the Role of the EDI-Bill of Lading And its Role to Achieve Possible Fullest

A. Elentably
Maritime Economics, King Abdul-Aziz University, Kingdom of Saudi Arabia

ABSTRACT: With a steady increase in maritime traffic of foreign trade of the world, and the continuing trend to maximize returns for investors and states alike, the time factor in the flow of goods linchpin of achieving those savings, and then highlight the absolute importance of handling freight bill, a title of goods of various types, and to achieve electronic exchange of invoice is on the top priorities for the departments of marine ports to achieve those savings, since the bill of lading for goods as a title to the goods traded electronically requires a thorough understanding of certain aspects which are complementary to each other, such as: Rules of procedure When not in conflict with these Rules, the Uniform Rules of Conduct for interchange of Trade Data by Teletransmission, 1987 (UNCID) shall govern the conduct between the parties. .beside that the Form and content of the receipt message and its meaning The carrier, upon receiving the goods from the shipper, shall give notice of the receipt of the goods to the shipper by a message at the electronic address specified by the shipper. In addition This receipt message shall include different information such as: the name of the shipper; the description of the goods, with any representations and reservations, in the same tenor as would be required if a paper bill of lading were issued; the date and place of the receipt of the goods; and a reference to the carrier's terms and conditions of carriage; plus the Private Key to be used in subsequent Transmissions, also the role of Terms and conditions of the Contract of Carriage , Right of Control and Transfer, the terms of Delivery, Option to receive a paper document , Through those tangles and measures to deal electronically bill of lading, highlights the importance of the element of time to achieve the fullest possible use of electronic exchange of data bill of lading and the implications of this exchange to achieve a standard rate of loading and unloading and to reduce waiting times for ships in ports to the marine and rates of performance standard berths port and optimize the use of journals and equipment docks and achieve financial savings from shipping operations. These rules shall apply whenever the parties so agree.

1 DEFINITIONS

– "Contract of Carriage" - means any agreement to carry goods wholly or partly by sea.
– "EDI" means Electronic Data Interchange, i.e. the interchange of trade data effected by *teletransmission.*
– "UN/EDIFACT" - means the United Nations Rules for Electronic Data Interchange for Administration, Commerce and Transport.
– "Transmission" - means one or more messages electronically sent together as one unit of dispatch which includes heading and terminating data.
– "Confirmation" - means a Transmission which advises that the content of a Transmission appears to be complete and correct, without prejudice to any subsequent consideration or action that the content may warrant.
– "Private Key" - means any technically appropriate form, such as a combination of numbers and/or letters, which the parties may agree for securing the authenticity and integrity of a Transmission.
– "Holder" - means the party who is entitled to the rights.
– "Electronic Monitoring System" - means the device by which a computer system can be examined for the transactions that it recorded, such as a Trade Data Log or an Audit Trail.
– "Electronic Storage" - means any temporary, intermediate or permanent storage of electronic data including the primary and the back-up storage of such data.

2 HOW CAN HANDLING OF PROCEDURE

When not in conflict with these Rules, the Uniform Rules of Conduct for interchange of Trade Data by Teletransmission, 1987 (UNCID) shall govern the conduct between the parties. The EDI under these Rules should conform with the relevant UN/EDIFACT standards. However, the parties may use any other method of trade data interchange acceptable to all of the users. Unless otherwise agreed, the document format for the Contract of Carriage shall conform to the UN Layout Key or compatible national standard for bills of lading. In the event of a dispute arising between the parties as to the data actually transmitted, an Electronic Monitoring System may be used to verify the data received. Data concerning other transactions not related to the data in dispute are to be considered as trade secrets and thus not available for examination. If such data are unavoidably revealed as part of the examination of the Electronic Monitoring System, they must be treated as confidential and not released to any outside party or used for any other purpose. Any transfer of rights to the goods shall be considered to be private information, and shall not be released to any outside party not connected to the transport or clearance of the goods.

2.1 *Form and content of the receipt message*

1 The carrier, upon receiving the goods from the shipper, shall give notice of the receipt of the goods to the shipper by a message at the electronic address specified by the shipper.
This receipt message shall include: the name of the shipper; the description of the goods, with any representations and reservations, in the same tenor as would be required if a paper bill of lading were issued; the date and place of the receipt of the goods; a reference to the carrier's terms and conditions of carriage; and the Private Key to be used in subsequent Transmissions. The shipper must confirm this receipt message to the carrier, upon which Confirmation the shipper shall be the Holder.
2 Upon demand of the Holder, the receipt message shall be updated with the date and place of shipment as soon as the goods have been loaded on board.
3 The information contained in paragraph (b) above including the date and place of shipment if updated in accordance with paragraph (c) of this Rule, shall have the same force and effect as if the receipt message were contained in a paper bill of lading

2.2 *Terms and conditions of the Contract of Carriage*

1 It is agreed and understood that whenever the carrier makes a reference to its terms and conditions of carriage, these terms and conditions shall form part of the Contract of Carriage.

2 Such terms and conditions must be readily available to the parties to the Contract of Carriage.
3 In the event of any conflict or inconsistency between such terms and conditions and these Rules, these Rules shall prevail.

2.3 *Applicable Law*

The Contract of Carriage shall be subject to any international convention or national law which would have been compulsorily applicable if a paper bill of lading had been issued.

3 THE SELECTION RIGHT OF CONTROL AND TRANSFER

The Holder is the only party who may, as against the carrier: claim delivery of the goods; nominate the consignee or substitute a nominated consignee for any other party, including itself; transfer the Right of Control and Transfer to another party;
1 instruct the carrier on any other subject concerning the goods, in accordance with the terms and conditions of the Contract of Carriage, as if he were the holder of a paper bill of lading. A transfer of the Right of Control and Transfer shall be effected: by notification of the current Holder to the carrier of its intention to transfer its Right of Control and Transfer to a proposed new Holder, and confirmation by the carrier of such notification message, whereupon the carrier shall transmit the information as referred to (except for the Private Key) to the proposed new Holder, whereafter the proposed new Holder shall advise the carrier of its acceptance of the Right of Control and Transfer, whereupon the carrier shall cancel the current Private Key and issue a new Private Key to the new Holder.
2 If the proposed new Holder advises the carrier that it does not accept the Right of Control and Transfer or fails to advise the carrier of such acceptance within a reasonable time, the proposed transfer of the Right of Control and Transfer shall not take place. The carrier shall notify the current Holder accordingly and the current Private Key shall retain its validity.
The transfer of the Right of Control and Transfer in the manner described above shall have the same effects as the transfer of such rights under a paper bill of lading.

4 THE PRIVATE KEY

1 The Private Key is unique to each successive Holder. It is not transferable by the Holder. The carrier and the Holder shall each maintain the security of the Private Key.

2 The carrier shall only be obliged to send a Confirmation of an electronic message to the last Holder to whom it issued a Private Key, when such Holder secures the Transmission containing such electronic message by the use of the Private Key.

3 The Private Key must be separate and distinct from any means used to identify the Contract of Carriage, and any security password or identification used to access the computer network.

5 DELIVERY

1 The carrier shall notify the Holder of the place and date of intended delivery of the goods. Upon such notification the Holder has a duty to nominate a consignee and to give adequate delivery instructions to the carrier with verification by the Private Key. In the absence of such nomination, the Holder will be deemed to be the consignee.

2 The carrier shall deliver the goods to the consignee upon production of proper identification in accordance with the delivery instructions specified in paragraph (a) above; such delivery shall automatically cancel the Private Key.

3 The carrier shall be under no liability for misdelivery if it can prove that it exercised reasonable care to ascertain that the party who claimed to be the consignee was in fact that party.

6 OPTION AVAILABLE TO RECEIVE A PAPER DOCUMENT

1 The Holder has the option at any time prior to delivery of the goods to demand from the carrier a paper bill of lading. Such document shall be made available at a location to be determined by the Holder, provided that no carrier shall be obliged to make such document available at a place where it has no facilities and in such instance the carrier shall only be obliged to make the document available at the facility nearest to the location determined by the Holder. The carrier shall not be responsible for delays in delivering the goods resulting from the Holder exercising the above option.

2 The carrier has the option at any time prior to delivery of the goods to issue to the Holder a paper bill of lading unless the exercise of such option could result in undue delay or disrupts the delivery of the goods.

3 A bill of lading issue shall include:
the information set out in the receipt message referred to (except for the Private Key); and a statement to the effect that the bill of lading has been issued upon termination of the procedures for EDI under the CMI Rules for Electronic Bills of Lading.

The aforementioned bill of lading shall be issued at the option of the Holder either to the order of the Holder whose name for this purpose shall then be inserted in the bill of lading or to bearer.

4 The issuance of a paper bill of lading shall cancel the Private Key and terminate the procedures for EDI under these Rules. Termination of these procedures by the Holder or the carrier will not relieve any of the parties to the Contract of Carriage of their rights, obligations or liabilities while performing under the present Rules nor of their rights, obligations or liabilities under the contract of carriage.

5 The Holder may demand at any time the issuance of a print-out of the receipt message (except for the Private Key) marked as non-negotiable copy. The issuance of such a print-out shall not cancel the Private Key nor terminate the procedures for EDI.

7 ELECTRONIC DATA IS EQUIVALENT TO WRITING

The carrier and the shipper and all subsequent parties utilizing these procedures agreed that any national or local law, custom or practice requiring the Contract of Carriage to be evidenced in writing and signed, is satisfied by the transmitted and confirmed electronic data residing on computer data storage media displayable in human language on a video screen or as printed out by a computer. In agreeing to adopt these Rules, the parties shall be taken to have agreed not to raise the defense that this contract is not in writing.

7.1 *INTRODUCTION*

This paper aims to give some idea of the **dynamics involved in implementing electronic bills of lading**. The bill of lading is one of the compendium of documents used in carriage of goods by sea. The writer did therefore not attempt to isolate the bill of lading, although the emphasis is clearly placed on substituting the traditional (*tangible*) bill of lading with EDI.

To understand the complexity of adapting existing documentation to EDI, it is essential to place the bill of lading into an EDI context. The electronic transfer of documents is nothing new. It is the statutory requirements and legal rights and obligations associated with the transfer that is currently stretching the boundaries of the law. Most of the legislation dealing with carriage and shipping documentation was drafted in an age where EDI was clearly not envisaged. Consequently, uncertainty exists regarding the legal recognition of electronic documentation.

The **role and function of the traditional bill of lading** is briefly examined followed by the electronic evolution of the bill of lading.

The **technical and legal obstacles** to the implementation of EDI are then reviewed. These include the requirement that the document has to be in writing, signature, negotiability and liability. The admission of computer generated evidence is also dealt with.

Parties wishing to enter the arena of electronic documentation will have to draw up an **interchange agreement** to regulate the various technical and legal issues arising out of the electronic transfer of documents. Various model interchange agreements are examined. The interchange agreement will in many ways be the backbone of the EDI operation. Parties will have to consider the legal and technical issues that might arise in the interchange agreement. A properly drafted interchange agreement will go a long way towards reducing some of the potential problems associated with electronic transactions.

EDI model rules provide for the incorporation of EDI into an acceptable legal framework. These rules will be considered. The emphasis is placed on the **CMI** Model Rules. The **UNCITRAL** Model Law on Electronic Commerce will also be briefly examined. This model law should provide a great impetus towards EDI acceptance and full scale legal recognition.

The paper then focuses on the attempt by various bodies to implement electronic bills of lading. Two prominent examples are given namely **Bolero** and **SeaDocs**. Lastly, a brief introduction is given to the impact of the Internet on EDI. This is an existing development and deserves further discussion. In conclusion, it is suggested that the traditional bill of lading can be substituted by EDI.

7.2 THE ordinary BILL OF LADING

The use of the bill of lading is almost as old as maritime trade itself. One of the earliest references to the keeping of records for cargo shipped on board is found in *The Ordonnance Maritime of Trani* of 1063. The original function of the bill of lading was therefore to acknowledge that the goods have been shipped. The use of the bill of lading became widespread during the 16th century and continued to develop as a respected document in international trade. Growing trade eventually necessitated the transfer of title in the goods before they arrived. It therefore became necessary to endorse the bill of lading to a third party in order to effect transfer of the goods. The bill of lading became a negotiable instrument. *Mitchelhill* reports that the first reported case in which endorsement of the bill of lading is mentioned dates from 1793.

The importance of the traditional bill of lading in international trade is largely self-evident when viewed against its functions. It is:
- Evidence of the contract of affreightment i.e. it contains all the essential terms;
- Prima facie evidence of the receipt issued by the carrier that the goods have been shipped or are received for shipment; and
- A 'quasi negotiable' document which passes the title in the goods.

International traders will almost always enter into a contract of carriage before the bill of lading is issued. The contract of carriage is then evidenced by the bill of lading. It is only possible to exclude this provision by express agreement.4 Furthermore, the bill of lading will also normally contain the terms of the contract of carriage.

Arguably, the most important function of the bill of lading relates to its negotiability. The bill of lading serves as negotiable commercial paper thereby enabling the transfer of title of the goods while they are in transit. Under English law, the bill of lading is not a truly 'negotiable' instrument because the indorse of the bill of lading can not receive a better title than the original holder had.5 However, the bill of lading is a document of title and the holder of the bill of lading is entitled to take delivery of the goods. This is settled law and is reflected in a 1912 House of Lords 6 decision where it was held that:

delivery of the bill of lading when the goods are at sea can be treated as delivery of the goods themselves, this law being so old that I think it quite unnecessary to refer authority for it.

The fact that the bill of lading is a document of title presents one of the greatest obstacles to the implementation of the electronic bill of lading. The effect will be examined later in this paper.

The traditional bill of lading also has several **disadvantages** in the modern shipping environment. Containerization and modern vessels have resulted in a speedier carriage of goods. The result is that the goods arrive at the port of destination before the relevant shipping documents. This causes delay and erodes the advantage gained by the expedited voyage. Considerable expenses are also incurred in the issuing and processing of bills of lading.7

The issuing of fraudulent bills of lading has also become a matter of international concern. Bills of lading are customarily issued in sets of three, consequently there is scope for the fraudulent use of more than one original to sell cargo on the water. These bills of lading are falsified in a number of ways:8
- Altering the quantity and quality of goods shipped in the bill of lading;
- in spite of the fact that an original bill of lading has been issued, the consignor may fraudulently sell the goods to other buyers during transit;
- the bill of lading can also be counterfeited in order to obtain fraudulent delivery; and

– it is possible to forge the bill of lading in order to obtain payment in a documentary credit.

There are various kinds of bills of lading. The form of the bill of lading will depend on the required function. Mitchelhill 9 lists the following types of documents:
– Bill of lading issued with printed clauses for conventional or through traffic on liner terms;
– Bills of lading issued for goods accepted under 'Combined Transport' conditions;
– 'Short form' or 'blank form' bills of lading;
– Bills of lading issued under a charter party; and
– Bills of lading issued by a freight forwarder.

Negotiable bills of lading are not always required. The result is that there has been an increase in recent years in substitutes for the traditional bill of lading. One such example is the **sea waybill**. Unlike the bill of lading, the sea waybill is not a document of title. It is intended for use where there is no transfer of goods envisaged. The sea waybill constitutes evidence of the receipt of the goods by the carrier as well as the contract of carriage. It is not necessary for the consignee to produce the document in order to obtain delivery of the goods. The consignee would merely have to produce adequate identification.10 This document is however not without inherent risks. The buyer who has paid in advance might find that the seller has changed the identity of the consignee. It therefore does not offer the same level of security that a traditional negotiable bill of lading does.

7.3 THE ELECTRONIC PROGRESSION OF THE BILL OF LADING

The traditional bill of lading has evolved over time to reflect commercial realities. Maritime commerce has been at the forefront of commercial development since its inception. It is therefore not surprising that the shipping industry has embraced the concept of an electronic bill of lading. Attempts are now afoot to replace the traditional, tangible bill of lading with electronic data.

Containerization has been the catalyst in introducing electronic data interchange to shipping documentation.11 Shipping and cargo interests started competing in an increasingly competitive environment. Computers made it possible for shipping documents to be processed quicker and more effectively than the traditional paper based documentation. It was therefore only a matter of time before electronic documentation was introduced in the shipping market. An EDI system would enable the parties to reduce the volume of documentation and the delay caused in transferring the documents.

Yiannopoulos 12 provides a strikingly accurate comment when he reflects on the development of the electronic bill of lading. He suggests that:

The [electronic bill of lading] is not a mere evolution in the form of bills of lading; it is the creation of a new species of bills of lading.

The bill of lading is issued by or on behalf of the carrier after the goods have been loaded on board. The holder of the bill of lading is therefore entitled in law to ownership of the goods. However, it can be endorsed to a 3rd party who then becomes the legal holder of the bill of lading and is entitled to take delivery of the goods. It is this transferability of the document that presents the real challenge to develop an EDI system for negotiable bills of lading.13

Besides, that can even take advantage and to achieve savings from the use of electronic data, and ensure the achievement of those savings must be adapted to many organizations and operational organizations for the inclusion of a number of laws and regulations should be binding for the introduction of the application of electronic exchange and work to ensure the rights of the Parties to maritime transport, either the shipper or carrier, or owners of the goods And produce a variety of laws that are binding on the authorities of ports and marine transport for the adoption of multi-electronic bills of lading And work to reduce the number of paper documents which used and also to reduce the duration of the cargo handling operations, which reflected positively on the parties to the process of maritime transport and leads to increases in growing the added value of Maritime Transport Sector

The impact of EDI on the traditional bill of lading also has to be evaluated against the formal requirements for a valid bill of lading. Most shipping nations subscribe to the Hague-Visby rules. Most of the international rules applicable to bills of lading were codified in the Hague Convention. This Convention was later amended by the Visby-protocol and became known as the Hague-Visby rules. No express provision is made in these rules regarding the formalities of a bill of lading. The Hague-Visby rules are applicable to any bill of lading relating to the carriage of goods. These rules are set out in the Schedule to the South African COGSA and have force of law in South Africa. Article III(3) reads:

After receiving the goods into his charge the carrier or the master or agent of the carrier shall, on demand of the shipper, issue to the shipper a bill of lading...

The implication is therefore that a document has to be issued. Will EDI satisfy this requirement? There is no specific reference to the fact that these documents have to be in writing or on paper. It is furthermore important to inquire into the formalities prescribed by each local forum. In Germany, for example, a bill of lading without a hand-written signature is null.14 Many other national laws and domestic legislation requires the use of paper documents. These requirements, which could present a serious obstacle to the use of EDI, will be dealt with later in this paper.

In addition to the legal obstacles involved in implementing EDI, several other factors also have to be taken into consideration. These include the technical aspects involved in setting up an EDI network. In order for EDI to be used effectively, it has to provide a secure means of transmitting the information. The trading partners will have to feel confident that the electronic messages are private and provide adequate protection against fraudulent misuse.

It is clear that the implementation of the electronic bill of lading holds many challenges. The success of this new species of bills of lading will depend on the work and effort of all the interested parties. Ultimately however, a wide scale acceptance will depend on practical results.

If the electronic bill of lading suits to the needs of the modern shipping industry and amplifies the functions of the traditional bill of lading, it will secure its survival in the competitive shipping environment.

7.4 *EDI*

The benefits of electronic commerce are widely accepted. Electronic Document Interchange (EDI) in particular has evoked considerable interest in recent years. EDI has been developed to allow computers to copy the relevant elements of data from a preexisting source within a subsequent message, thereby eliminating re-keying and duplication of activities.15

In order to understand the impact of EDI on international trade and commercial transactions, it is necessary to examine how EDI functions. A number of important legal considerations also arise in the process of facilitating electronic commerce. These considerations have been alluded to above, and will be discussed in more detail.

7.5 *UNDERSTANDING EDI*

A number of definitions has been formulated for EDI. In essence, EDI is:

*...the replacement of the paper documents relative to an administrative, commercial, transport or other business transaction, by an electronic message structured to an agreed standard and passed from one computer to another without manual intervention.*16

EDI, as a means of conducting business, is gaining popularity and acceptance for a number of reasons. These are:

- EDI increases the speed with which business is conducted by eliminating the delay caused by manual (paper-based) documentation. The transfer of documentation is therefore speeded up. It would eliminate the delay caused by cargo arriving at the port of destination before the actual documentation required to take delivery arrives.
- Messages sent by EDI are also accurate since the information is structured to an agreed format. The result is that the message will be rejected if it does not conform to this format. The electronic information would furthermore be verifiable by means of either a 'private key' or electronic signature.
- Digital encryption ensures that the message is authentic. Fraud would therefore be reduced.
- All of these factors ensure that a company trading via EDI would save time and money.

In spite of all these advantages, EDI is not as extensively used as one would expect. One of the reasons for this is the legal uncertainty surrounding EDI. The capital expense involved has also inhibited the development of EDI in some sectors. In spite of this, EDI is poised to have a significant impact on the way business is conducted in the immediate future.

An EDI message consists of several parts. The messages approved for EDI use also have to be incorporated into a message framework. This would then in effect provide *"a language of alpha-numeric codes around which the content of each EDI message is constructed and a 'grammatical' structure through which those codes can be organised"*.This message can then be divided into three concentric subparts:

- The first subpart is the message itself, e.g. the electronic version of the bill of lading.
- The transaction set is then made up of segments.
- The segment is then made up of data elements.

The message framework and code list will then translate these conventional terms into a computer understandable unit. This implies that the EDI users will not have to enter a complete new set of data/information into the computer every time a new transaction is conducted.

In order for EDI to function there has to be a combination of technology and management resources to ensure that the data is transmitted correctly and accurately between the computer systems of the various parties. Parties need to ensure that the correct software is utilised to transmit the internal data format to an acceptable EDI format. It might also be required to make use of a third party EDI network.

8 THE TRANSACTION OF VALUE-ADDED NETWORKS (VAN)

Although it is possible for EDI users to link their computer systems directly to each other, in practice they would often make use of a Value-Added Network service provider (VAN). These firms specialize in technical assistance. VAN's would also provide technical support and assist in data security and the configuration of the required software.

Computer programmes and data systems are not always compatible with each other. For example: the carrier could use a computer system which cannot process the information received by the computer system of the shipper. In such a scenario, the carrier and the shipper would make use of a VAN. The biggest advantage of a Value Added Network service provider is the fact that it can bridge the gap between these two systems. In other words, the VAN will match the various computer protocols and provide the necessary software. This ensures that data that has been created on one system can be received on the other system.

Most of these networks operate on a generic basis. The network will offer its services to any party entering into an agreement with it. However, it is also possible for networks to specialise and provide their services only to a particular class of users. It would therefore be possible for a network to specialise in the movement of EDI information and documentation exclusive to the shipping industry. In practice however, it is likely that the parties will have to make use of multiple networks because of the vast amount of documentation involved originating from the various sectors in an international trade transaction. In some cases, an additional network might be required to connect the different parties to each other.

The choice of the Value Added Network is crucial to the operation of the EDI transmission. The VAN will control the communication between the various parties and will hence be responsible for the smooth operation of the electronic transfer of the relevant documentation. If the VAN experiences problems or shuts down completely, this will directly affect the transfer of the electronic documentation. Liability issues are also likely to arise under these circumstances and these issues will have to be regulated in an underlying user agreement.

9 SOLVE OF TECHNICAL PROBLEMS

There are a number of technical problems associated with EDI. One of these is the fact that electronic documents have to be exchanged according to a certain common standard. A standard data format would therefore be required in order to ensure compatibility between the various systems currently in use. Furthermore, it is essential for the relevant data documentation to conform to adequate security standards. Opponents of EDI argue that the electronic transmission of data is not secure enough to provide a solid foundation for transmitting the bill of lading on a computer. These are valid concerns and will have to be addressed. However, it is submitted that various techniques exist to secure the data

transmission and provide for the integrity of the message. These techniques include encryption and the use of Personal Identification Numbers. Alleviating fears about real and perceived lack of security will be a great challenge to the proponents of EDI.

9.1 *Useful STANDARDS*

Electronic documents have to be exchanged in a standard data format. The computer has to process the data to enable the data to become information that can be understood by the receiver. Document content standards are used for this purpose. These standards will then ensure that the order in which data appears is fixed to a certain common standard.

Unfortunately, the search for a common standard has resulted in two very different standards being developed. In the United States of America the ANSI X12 cross-industry standard (American National Standards Institute Accredited Standards Committee X.12) is widely in use while the United Nations (in co-operation with the International Standards Organization) has developed EDIFACT (EDI for Administration, Commerce and Trade). EDIFACT consists of:

a set of internationally agreed standards, directories and guidelines for the electronic interchange of structured data, and in particular that relating to trade in goods and services, between independent computerised information systems.

The EDIFACT language is made up of a comprehensive coded data register. This register basically covers all the words and printed forms used in trade. Furthermore; it provides a common syntax and format that will result in the production of recognizable shipping documents. It would not make a difference if the hardware and the software used are not compatible. Human intervention will therefore not be necessary to process the information. This will enable the bulk of the shipping documentation to be processed at a speedy rate which in turn ensures efficiency and savings in costs. A carrier could therefore send a computerized bill of lading according to an agreed standard (e.g. EDIFACT); the shipper's computer will instantly recognise the document as a bill of lading and proceed to conduct computer operations on the document.

The lack of a universally agreed standard should not necessarily be seen as a bar to the growth of international trade or the use of EDI. Some commentators have suggested that a common standard is not desirable in an industry where every sector has its own unique way of communicating and conducting business.

In order to ensure the full benefit of a universally recognised standard, it is suggested that parties should specify the standard to be used in the interchange agreement. This would ensure that an added degree of interchange security is obtained. The tech-

nical problems relating to a common 'language' or agreed standard can therefore easily be overcome by co-operation of the various sectors involved in the exchange of information by EDI. Other problems however remain, these are:
- Providing the necessary hardware and software service. These services will have to be agreed upon in the interchange agreement.
- Providing adequate backup procedures for emergencies. It is essential to establish the liabilities involved in the event of a communications breakdown.

It is suggested that these technical problems can easily be overcome by drafting a proper interchange agreement to regulate the EDI operations of the parties. However, the greatest area of concern for the parties will be to provide for adequate security.

9.2 Requirement's SECURITY

In order for traders to be comfortable with the use of EDI, they will have to be satisfied that the system as a whole and the message in particular, is secure. Security weaknesses will also inhibit the legal acceptance of EDI transactions. Various methods are used to ensure that the electronic data is transmitted on a secure basis. These methods include passwords, encryption, PIN codes and electronic signatures. Encryption will ensure that the data transmission is kept confidential while authentication will provide for data integrity.

9.3 The Method of Protecting the System

The data stored in the computer system is susceptible to tampering. Access to the data will therefore in most cases be restricted to authorised users. The use of an access card is one way of ensuring that only an authorised person uses the system. It is almost impossible (and neither financially viable) to provide a foolproof system. The parties would therefore have to agree to the level of security needed in order to minimize the risk of fraud or tampering. The need for security has also been recognized by UNCITRAL, stating that:

...it is cleat that the legal reliability of EDI techniques requires that high standards be used to determine legal certainty as to the identity of the sender, its level of authorisation and the integrity of the message.

It might also be useful for parties to have their system audited by a security expert at various intervals. The security expert should be independent and must ensure that the required security measures have been implemented. This would provide for an added sense of security between the parties.

9.4 Protecting the Integrity of the Message

The message integrity can be assured by the process of authentication, i.e. ensuring that the data sent has not been tampered with. Encryption would seem to provide the most security but it must be noted that encryption techniques are prohibited by some governments as reported in a recent Time magazine article. Governments opposing the export of encryption techniques fear that this technology might be abused by criminals and terrorists. Parties would normally agree to the use of encryption in the interchange agreement. One example of a model interchange agreement that provides for encryption is UNCID. Article 9(b) deals with the possibility of parties to agree to use encryption.

An interesting development concerning EDI security has been the approval of a resolution by the American Bar Association (ABA) dealing with legal-security issues involving electronic data interchange and electronic commerce. According to this resolution, the ABA has to:
- facilitate and promote the orderly development of legal standards to encourage use of information in electronic form, including appropriate legal and professional education;
- encourage the use of appropriate and properly implemented security techniques, procedures and practices to assure the authenticity and integrity of information in electronic form; and
- recognise that information in electronic form, where appropriate, may be considered to satisfy legal requirements regarding a writing or signature to the same extent as information on paper or in other conventional forms when appropriate security techniques, practices, and procedures have been adopted.

As has already been mentioned, cryptography offers a viable means of providing security. However, the costs of implementing these measures are often quoted as an inhibitor. This problem was addressed in a workshop (conducted by the National Institute of Standards and Technology of Gaithersburg, Maryland) on Security Procedures for the Interchange of Electronic Commerce. The cost in implementing cryptographic methods would include software licensing, export filing process, overheads and professional training of staff. It was argued that a premature consideration of costs could eliminate other viable options. The parties would therefore have to evaluate their underlying requirements to determine what level of security is required. Since a bill of lading is a document of title and entitles the holder to claim delivery of the goods, the level of security needed would have to be substantially higher than the security required for a normal receipt.

Security services will have the added benefit of providing services that are not possible to provide with paper-based techniques. An example of such a service is non-repudiation. This method ensures that the originator of a document cannot deny the origin of the document, thereby providing irrevocable proof of authenticity.

Digital signatures are limited in some respects. Parties making use of EDI or digital signatures will often have to revert to a trusted third party to provide security assurance. The third party will be required to date-stamp, store and keep an audited data log of the transaction. This would provide proof of the time of origination and content of the electronic document. Once again, the liability issues arising from the use of a third party will have to be worked out in the interchange agreement.

Proper message or data authentication will also enhance the evidential value of the message. The court will have to be sure that the message submitted as evidence is authentic. Admissibility of EDI evidence will be dealt with later.

10 LEGAL PROBLEMS

EDI has been the catalyst for a number of changes in the scope and function of the law. Legal reform has however not always kept pace with technological development. The legal problems involved in implementing EDI on a global basis become apparent when viewed against the relevant statutory requirements imposed by the various jurisdictions. Bills of lading have to meet certain statutory and formal requirements before they become legally enforceable. These requirements will now be examined.

11 CONCLUSION

It is difficult for parties to relinquish a document which has served them well over the years. The bill of lading is one of the most respected documents in international trade. It has been held that:

*A bill of lading is a document of dignity, and courts should do everything in their power to preserve its integrity in international law for there, especially, confidence is of the essence.*179

However, new technology has brought new challenges and possibilities. Once the remaining technical concerns regarding security and authentication have been resolved, and legal recognition assured, full scale implementation is possible.

Substituting the traditional bill of lading with EDI is still fraught with real (and perceived) problems. Parties wishing to trade with EDI will have to be aware of the potential pitfalls associated with electronic trading. Pending legislative reform, the parties will have to regulate many of the technical and legal requirements in the underlying EDI interchange agreement.

It is possible even take advantage and to achieve savings from the use of electronic data, and ensure the achievement of those savings must be adapted to many organizations and operational organizations for the inclusion of a number of laws and regulations should be binding for the introduction of the application of electronic exchange and work to ensure the rights of the Parties to maritime transport, either the shipper or carrier, or owners of the goods And produce a variety of laws that are binding on the authorities of ports and marine transport for the adoption of multi-electronic bills of lading And work to reduce the number of paper documents which used and also to reduce the duration of the cargo handling operations, which reflected positively on the parties to the process of maritime transport and leads to increases in growing the added value of Maritime Transport Sector.

The traditional bill of lading will still have its place in the immediate future. The capital expenses of setting up an EDI network might prove too costly to afford these services to everybody. There is no reason why the electronic bill of lading can't co-exist with the paper bill of lading for the immediate future. The real challenge lies in creating a system in which both traders and the courts feel comfortable. This will require a concerted global effort from all the parties involved. This is certainly a difficult, but by no means impossible, task. The proponents of EDI will have to prove that the electronic bill of lading can function in the real trading environment and, ultimately, that it provides users with a competitive edge.

The electronic bill of lading will undoubtedly become a reality. There are simply too many advantages attached to this form of trading to dismiss the concept. In order for the electronic bill of lading to replace the paper bill of lading, it would essentially have to offer the same advantages and level of security associated with the paper bill of lading.

The functions of the negotiable paper bill of lading can be duplicated, but the electronic bill of lading would have to go one step further: it would have to improve on the traditional bill of lading. The advantages of the electronic bill of lading have been discussed. It is suggested that these advantages will prove sufficient to eventually replace the paper bill of lading and take the bill of lading into the next century.

REFERENCES

[1] Bainbridge D. Introduction to Computer Law 3rd Ed. Pitman Publishing 1996
[2] De Boer Th.M. The Missing Link: Some thoughts on the relationship between private international law and comparative law in Boele-Woelki et al Martinus Nijhoff (Dordrecht) 1994
[3] Delatista, C. Incoterms in Practice ICC Publication 1995

[4] Farlam & Hathaway. Contract 3rd Ed. Juta & Co Ltd 1994

[5] Forsyth, C.F. Private International Law 2nd Ed. Juta (Cape Town) 1990

[6] Hinge, K.C. Electronic Data Interchange -- From Understanding to Implementation AMA Membership Publication Division (New York) 1988

[7] Mitchelhill, A. Bills of Lading Chapman & Hall 1982

[8] Payne & Ivamy Carriage of Goods by Sea 13th Ed. Butterworths (London) 1989

[9] Scrutton on Charterparties 19th Ed. Sweet & Maxwell (London) 1984

[10] Walden, EDI and the Law Blenheim Online Publishing (London) 1989

[11] Wheble, B The EDI Handbook Blenheim Online Publishing (London) 1988

[12] Yiannopoulus, A.N. Ocean Bills of Lading: Traditional Forms, Substitutes, and EDI Systems Kluwer Law International 1995

[13] Chandler G.F. "Maritime Electronic Commerce for the Twenty-First Century" Paper presented to CMI Panel on EDI June 10th 1997 at Antwerp, Belgium

[14] Debatista C. "Incoterms in Practice" ICC Publication 1995

[15] Eiselen G.T.S. "Elektroniese Dataverwisseling en die Bewysreg" THRHR 1992(5)

[16] Eiselen S. "The Electronic Data Interchange Agreement" SA Mercantile Law Journal 7 1995

Maritime Transport Policy

12. Effectiveness of the European Maritime Policy Instruments

G. Šteinerts
Latvian Maritime Academy, Riga, Latvia

ABSTRACT: European Maritime Transport Policy 2009-2018 was confirmed by EC Strategy Paper of 21 January 2009. This gives sound basis also for future of European shipping policy and particularly in respect of the global competitive position of European shipping. State Aid Guidelines for Maritime Transport have been introduced and implemented in many European countries and it was proposed now to maintain these Guidelines for another longer period and possibly to improve them also. The latest EU Maritime policy developments are discussed more in detail in this Paper and effectiveness of them assessed.

1 EUROPEAN LEGISLATION ON STATE AID

1.1 *State Aid Guidelines for Maritime Transport*

Basic principle of European State aid to maritime transport is based on gains from maritime cluster that overweight the tax relaxation and reduction of State taxation revenues from the shipping as a consequence.

Initially the State aid to the shipping industry was introduced since 1989, based on the example of Greece and the Netherlands. Now the European Shipping policy is mainly based on Communication of the European Commission C(2004)43 of 17 January 2004 "Community Guidelines on state aid to maritime transport" [1]. The term of application of these Guidelines was extended up to year 2011 at least. These Guidelines are focusing on merchant fleet taxation system, introducing tonnage tax instead of corporative tax, also relaxation of income taxation and social security payments for community seafarers. Many European countries had introduced this State aid to merchant shipping, they have to report to European Commission and obtain approval from EC. The above Guidelines are under revision as announced by COM(2009) 8, the Communication of 21 January 2009 "Strategic goals and recommendations for the EU's maritime transport policy until 2018". COM(2009) 8 strongly declares prolongation of Guidelines: "A clear and competitive EU framework for tonnage taxation, income taxation and state aid should be maintained and, where appropriate, improved, in the light of the experience gained under the State aid guidelines for maritime transport." [2]

Apart from the above type of State aid to ship owners and seafarers taxation relaxation in Chapter 10 of Communication C(2004)43 is the provision providing another type of State aid in a form of complementary funding for the launching of the Motorways of the Sea. This is confirmed by another Communication 2008/C 317/08. By this communication is allowed, under certain conditions, for start-up aid to new or improved short sea shipping services with a maximum duration of three years and a maximum intensity of 30 % of operational cost and 10 % of investments costs. This is Marco Polo scheme, focusing on financial support by Community funding in development of new short sea shipping lines. Second stage of Marco Polo II program has been successful. Under the 2009 call for proposals, 22 projects were successful - from 70 bidders for the budget of €66.34 million. Marco Polo is supporting also rail transport, accounting for 40.7% of the grant, followed by sea routes with 22.7%. The project budget for 2007 – 2013 is €450 million. [3]

1.2 *State Aid to ship management companies*

This is new undertaking by EU to extend the Commission Communication C(2004) 43 provisions on reduction of corporate tax or the application of the tonnage tax also to ship management companies under section 3.1. of the Guidelines. This is provided by adoption of Communication from the European Commission 2009/C 132/06. It does not deal with the State aid to commercial managers of ships. The Communication applies to crew and technical management irrespectively of whether they are individually provided or jointly provided to the same ship.

However, eligibility is limited to the joint provision of both technical and crew management for a same vessel ("full management").

In Europe shipping management most developed is in Cyprus, which features the largest ship management industry in the world. Important players in this field are also in the United Kingdom, Germany, Denmark, Belgium and the Netherlands. Outside Europe ship management most developed is in Hong Kong, Singapore, India, United Arab Emirates and the USA. Therefore it is important that European ship management business should be further supported in respect of the global competitive position of European shipping be maintained and reinforced.

1.3 State Aid to shipbuilding

State aid provisions for Community shipbuilding have been introduced long before tonnage tax system. Present ones are given in the "Shipbuilding Framework" which entered into effect on 1 January 2004. [4] These rules have been applied initially for a period of three years, then prolonged twice and currently applicable until 31 December 2011. In a view of the expiry of the Framework by the end of 2011, the Commission have started consultations with Member States and other interested parties to determine whether to continue to apply this Framework, modify it or let it expire in 2011.

Some of the problems that the EU shipbuilding industry is facing today are linked to the economic and financial crisis. After a period of several years of high demand, in 2009/2010 demand for new ships has fallen drastically. Therefore this aid could be critical to support research, development and innovation in European shipbuilding as announced in COM(2009) 8, the Communication of 21 January 2009 "Strategic goals and recommendations for the EU maritime transport policy until 2018".

2 ECONOMIC BACKGROUND

2.1 European fleet position in the World

Europe plays a major role in today's shipping world, with European companies owning 41% of the world's total fleet in terms of deadweight (DWT) and 45% in terms of Gross Tonnage (GT). This is partly because of effectiveness of EU State aid regime. These figures are based on ownership, or to be more accurate, on real control of this fleet and not on flag.

European (European Economic Area – EEA) registered merchant fleet share in the World fleet, which may be used for assessment of EU State aid effectiveness, is as in Table 1. [6]

Table 1. European flags share in the World merchant fleet, mil. GT / DWT , %, as on 1st July 2010

	Gross Tonnage, mil.	Deadweight, mil.
World	915.976	1348.786
EEA	209.079	289.705
EEA / World %	22.8 %	21.5 %

As to latest statistics of European national flag registers share in the World, the situation is rather stable about 21- 22% share, but YoY development is slower than the Worlds' one (Fig.1). [6] [7] [8] [9] World merchant fleet is developing steadily and faster than the EAA fleet under national flags.

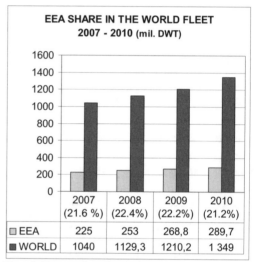

Figure 1. European flags share in the World merchant fleet, mil. DWT , %, 2007 – 2010

2.2 European registered and controlled fleet

The economical strengths of the European shipping are connected not only to national registers. It is always the matter of discussions whether the national flags are so important as far as the merchant fleet development very often is based of the Flags of Convenience (FOC) model and these flags are doing well indeed, the most successful being Panama with 198.6 mil. GT (21.7% of the World fleet), Liberia with 97.7 mil. GT (10.6%) and Marshal Islands with 55.7 mil. GT (6.1%).

At the same time even the 3 biggest European flags are comparatively small – Greece with 40.9 mil. GT (4.4% of the World), Malta – 35.7 mil. GT (3.9%) and Cyprus – 21.2 mil. GT (2.3%). [6]

Malta and Cyprus also could be considered as European FOC registers and this is possibly because of ship management businesses in these countries. Another European FOC may be attributed to Isle of Man register, which is outside the EEA jurisdiction but with close connections to UK shipping.

The situation for all European registered merchant fleet versus European controlled fleet but registered under 3rd flags is rather stable last years. EEA Registered merchant fleet in terms of GT is of the same size as EEA controlled fleet but registered under 3rd flags, possibly under FOC. At the same time it should be noted, that not all fleet registered in EEA registers is controlled by EEA shipowners. The situation for EEA controlled and EEA registered merchant fleet and these two types of European fleet against World Total fleet in terms of GT for years 2007 to 2010 are presented in Table 2. [6] [7] [8] [9]

Table 2. EEA registered and EEA controlled merchant fleet 2007 – 2010 (100 GT and above), mil. GT

	2007	2008	2009	2010
EEA Registered and controlled	140.5	151.6	163.0	175.7
EEA Registered, foreign controlled	21.2	30.0	30.8	33.3
EEA controlled, registered outside EEA	150.3	160.2	160.8	206.5
TOTAL EEA	312.0	341.6	354.6	415.5
TOTAL World	704.6	770.9	824.7	916.0
TOTAL EEA to TOTAL World, %	44.2 %	44.3 %	43.0 %	45.3%

EEA Registered and foreign controlled fleet is rather unclear model. We can only guess where these foreign owners are from. Possibly the reason is connected with the development of European ship management businesses, such are very active in Cyprus. Very possibly this is also in conjunction with the offshore registered companies, i.e. very well known and favourite model of tax–avoidance. Whatever the reasons for foreigners to register their ships in EEA registers, this is in line with the European shipping policy supported by new European initiative on State Aid to ship management companies. Contrary to this, European shipping policy is not favouring European controlled fleet registered outside EEA, which is not in line of strengthening of EEA countries maritime cluster but supporting FOC system rather.

2.3 *Impact of the World economic recession*

Demand for shipping and port services is derived from the demand for the products being shipped. Collapse in demand because of the global financial crisis has lead to a dramatic change in freight rates, which also collapsed in all sectors and are unlikely to begin their upward cycle until the supply and demand gap is closed. In this respect, the emerging markets of China and India have made a significant impact on demand side of the industry. As far as trade is concerned, there is still enormous potential for growth in countries such as Vietnam, Malaysia, Philippines, the Indian sub-continent, countries of Central Asia, Russia and the East European states,

Africa and South America. Despite of current massive overcapacity of tonnage the growth of China, India and others Asia countries are likely to help avoid overcapacity and depression in the shipping market.

Shipping industry analysts are rather optimistic on recovery of shipping market. As to World merchant fleet development, they quote the following YoY figures, as in Table.3. [5]

Table 3. Prospects of World merchant fleet development 2011

	2009	2010	2011
Tonnage demand	-2.9 %	12 %	6 %
Fleet growth	6.7 %	7 %	7 %
Utilization rate	81.1 %	85 %	84 %

In connection with future development of European fleet, we may examine the shipbuilding portfolio, as at May 2010, showing the following figures in terms of deadweight:

EEA shipowners have ordered merchant fleet of 174.8 mil. DWT, which is 32.4 % of the World shipbuilding portfolio of 540 mil. DWT. These figures does not include 64 passenger ships' ordered for EEA shipowners out of 192 passenger ships total orders in the World. This is a promising sign also for fleet renovation trend as the newbuiding portfolio is approximately 40 % of existing fleet size. [6]

It is hard to assess what impact could be expected because of application of European State Aid to shipbuilding regime. This State aid may be of good use for European shipyards which are key drivers for maritime innovations and therefore eligible to this State aid regime. The difficulties may arise because of Far East shipbuilders' enquiries to World Trade Organization (WTO) whether this Sate aid complies with the fair competition rules.

3 EEA COUNTRY PROFILE

3.1 *Major European flags*

There are 10 EEA flags in the list of World 30 biggest flags. No.1 in EEA is Greece, ranking 7th in the World. Then follows Malta, 8th in the World and Cyprus, 10th in the World. It should be noted again however, that these statistics are according to registration or flag and not according to the fleet under ownership or under control.

If we consider the second type of ranking, i.e., controlled fleet, then No.1 in the World is Japan with 176mil. DWT and No.2 - Greece with 175 mil. DWT. Germany with 105 mil. DWT is No.3 in the World by controlled fleet. The current list of 10 EEA major flags includes also UK, Norway, Italy, Denmark, France, Netherlands and Isle of Man (not in EEA). The development of European major merchant fleets by flag as from 2007 is presented in Table 4. [6]

Table 4. Development of major European flags 2007 - 2010, mil. DWT and percentage

	2007	2010	2007/2010 %
Greece	56 665.4	71 752.8	+ 26.6 %
Malta	40 480.5	58 337.2	+ 44.1 %
Cyprus	30 141.5	33 024.6	+ 9.5 %
Norway	22 491.5	22 687.8	+ 0.9 %
UK	12 437.8	19 245.8	+ 54.7 %
Italy	13 346.0	17 682.2	+ 32.5 %
Germany	13 856.0	17 660.2	+ 27.4 %
Isle of Man	13 730.0	15 486.5	+ 12.8 %
Denmark	10 602.9	13 319.5	+ 25.6 %
Netherlands	4 999.4	7 828.8	+ 56.6 %
France	7 343.0	8 019.0	+ 9.2 %

Tonnage increase as from 2007 is noted for all major European flags except Norway. The champions for increase speed are UK and Malta. Serious increase of tonnage is also for Netherlands, the basic initiator for introduction of tonnage tax. Malta's tonnage increase apparently is connected with the ship management business. All this is confirming the effectiveness of the system of State aid in general, when this aid is properly applied.

However, this is current situation of tonnage under national flags. When considering tonnage under control Greece is No.2 in the World with 175 mil. DWT. Germany is No.3 with 105 mil. DWT and have the fastest developing fleet with average annual growth of 16.2% but 83% of it is foreign registered. [10]

3.2 *European country profile*

It would be important to assess the situation in each of EAA country, what is a balance between the national flag and 3rd flag fleets. This would serve as the indication of how effective is the system of State aid, which is a basis for development of a maritime cluster in each country and in EEA as a whole.

Greece is the first to consider as No.1 in EEA and No.2 in the World by controlled tonnage. The Greek owned fleet under EU flags (including national Greek flag) accounts for 39.7% of the EU tonnage in terms of DWT, while under national Greek flag is 24.7% of the Greek owned fleet. Therefore, under non-European flags is 60.3% of the Greek owned fleet and this is not in line with the European shipping policy ideas. Further development of Greek fleet is expected, as by the end of December 2009, new-building orders by Greek interests (ships over 1,000 GT) amounted to 748 vessels and 64.9 million DWT. Under what flag this new fleet is going to be operated is hard to answer.

Generally speaking, the Greek fleet situation confirms the effectiveness of application of tonnage tax system introduced in Greece before the Netherlands.

There are 1,300 shipping companies now in Greece giving solid base for national maritime cluster providing employment directly or indirectly to 200,000 persons. The outstanding performance of the maritime sector in the context of World (and national) economic downturn in 2009 is a confirmation that the maritime sector was the only in Greece that did not create any unemployment. Positive trend also is noted on national efforts for the attraction of youngsters to the seafaring profession, resulting in increase of number of cadets for the Marine Academies by 50%. [6]

Malta as a second European flag with 58.3 mil. DWT fleet is developing very fast. As to fleet under control of Malta's shipowners, the figure is very small (59,000 DWT on 1st July 2009). That means that all Maltese registered fleet is foreign owned. Malta shipping policy is based on support of ship management businesses, supported also by national law. In 2010 the new regulations have been adopted which extend the tonnage tax regime to foreign flagged ships and to ship-management activities.

Cyprus with 33.0 mil. DWT fleet is another example of the largest ship management centre in Europe. Merchant fleet under control of Cyprus shipowners accounts for 25% of Cyprus flag registry (2009). Cyprus is also implementing improvements to their tonnage tax system by offering new additional tax incentives for Cyprus shipping companies.

Norway historically is claiming to be a shipping nation. Norwegian International Ship Registry (NIS) was introduced long time ago serving some support to Norwegian fleet. As to taxation, in 2007 the Parliament has adopted a new shipping tax scheme abolishing favourable 1996 taxation rules. In February 2010, the Norwegian Supreme Court concluded that 2007 rules are not complying with the Constitution, previous rules may be applied. Statistics of development of Norwegian merchant fleet reflects these political uncertainties: for period as from 2007 to 2010, tonnage is on stagnation, in 2010 being 22,6 mil. DWT. Fleet under Norwegian shipowners control is a double size as under Norwegian flag (including NIS); about 49% (2009) of the controlled fleet is registered outside Norway. [10]

United Kingdom is one of the pioneers for introduction of tonnage tax 10 years ago resulting in rise of tonnage under UK flag while there has also been some lack of certainty in some aspects of the tonnage tax and UK's fiscal regime in general. During period of 2007 – 2010, the increase of tonnage was by 54.7%. Since 2008, a dramatic slowdown in trade and a collapse in shipping freight rates in 2009 this growth have been held back. Sea transport is in the UK is No.3 in of top export earners and positive contributor to employment through the tonnage tax regime. The number of cadets in training has more than doubled in the last ten years. The merchant fleet under UK registry and UK shipowners control is stable, last year increased by 1.2% to 21.5 mil. DWT.

Statistics on UK shipping usually include Isle of Man (IoM) flag registry being some kind of British FOC. The tonnage under UK + IoM shipowners control is about 32.2 mil. DWT with average increase of 5.9%. About 90% of the controlled fleet is under national flags. [10]

Italy with 17, 7 mil. DWT merchant fleet is on a fast rise of 32.5% during 2007–2010. Italian approach to State aid is based on establishment of the Italian International Register. 91.3% of Italian owned ships are entered in the Italian International Register, while 8.1% are entered in the Ordinary Register. A small proportion, about 3.8% of the total, of the Italian owned tonnage flies a foreign flag. [6]

Germany with the third largest merchant fleet under control of their owners, launched an initiative for the registration of ships under the German flag. German merchant fleet since 2000 has four times increased in size with an average growth rate of approx. 14 % per year over last decade. In container shipping Germany is the World leader with over one third of the controlled container vessel fleet. This position will also be kept in the future as German shipowners have ordered about 900 vessels representing 32 mil. GT on order. 243 of these 900 ships are containerships. While there is a positive summary of the German shipping policy in recent years, German Government is critical and consider whether State aid in a form of tonnage tax regime be prolonged.

Denmark registry (including DIS) with 13.3 mil. DWT merchant fleet is developing rather fast, by 25.6% during 2007 – 2010. Denmark is the fifth in the world by controlled tonnage of more than 50 mil. DWT. About 60% of the controlled fleet is under national flag. [8] Denmark is supporting revision of the EU State Aid Guidelines for tonnage taxation, income taxation and other types of State aid. [6]

Netherlands merchant fleet under national flag is growing, in 2010 being 7.8 mil. DWT, increased by + 56.6 % during 2007 – 2010. But still the shipowners and ships are deserting the Netherlands out of the necessity of cutting costs. This is showing that the Netherlands as a basic initiator for application of tonnage tax have insufficient shipping taxation regime. The Netherlands government has acknowledged that flagging out and the departure of companies and shore management from the country will have negative consequences also for national seafarers and for nautical education. [6]

Belgian merchant fleet tonnage at 12.5 mil. DWT last years was on a slight rise, by 8.8% annually. 47% of the Belgian controlled fleet was registered in the Belgian register, 44% in other EU registers and, unlike other European controlled fleets, just a minority of 9% in open registers. [6]

French merchant fleet of 8 mil. DWT (including FIS) is slowly developing. France does not use the tonnage tax system; instead, at the end of 2009 the guarantee scheme for ship financing, like for other companies in France may be used. It is not a state aid, there is a cost for the shipowner.

Sweden is serving negative example of governmental shipping policy. Because of political decisions in April 2009, it was announced that there would be no tonnage tax during the government's present term of office. As a result, the flagging out is the only solution for shipping companies and this is reflected in statistics: only 33.7% out of Swedish controlled fleet of 7.2 mil. DWT is under national flag. [6]

Spanish shipping is using Canary Island Special Register as a second one. Foreign flag is dominant by 67.3% out of 4.5 mil. DWT fleet. At the same time, Spanish registered fleet is decreasing and foreign registered increasing. [7] This apparently is because of inefficient shipping policy of the Spanish government.

In Portugal shipowners with the fleet of 1.2 mil. DWT are using Conventional Register and Madeira International Register, 25% of ships are flagged out. They are in need for an effective shipping policy, similar to those that have been successfully implemented in other EU countries. [6]

Polish shipowners control a fleet of 2.6 mil. DWT of which only 144 thousand DWT fleet or 5.5% is under national flag. A Tonnage Tax Act was adopted in 2006, formally approved by the Commission in December 2009 and came into force on 1 January 2011. [6] It is a hope that this State aid regime will improve the situation but pending on success of negotiations with trade unions.

Finnish merchant fleet is on stagnation with 1.2 mil. DWT under national flag. Revised tonnage tax regime is underway accompanied by difficult trade union negotiations but may improve the situation.

Latvian merchant fleet under national flag is small and in further decrease notwithstanding the State aid regime has been introduced in 2001. Flagging out (90%) is usual approach, including for Malta flag. Shipowners claim that too formal approach from the State revenue office is to be blamed. Lithuanian merchant fleet under national flag is on decrease having 0.3 mil. DWT size in 2010, now struggling the recession but introduction of tonnage tax is underway. Estonian shipping companies did not manage to pass the tonnage taxation law in the Parliament but hope to make it in future. The fleet under national flag is of 0.168 mil. DWT, the rest of the Estonian controlled fleet is flagged out including four vessels registered under the Latvian flag, which appears to be more favourable.

Bulgaria with 0.9 mil. DWT fleet under national flag still encounter massive flagging out (92.7%)

notwithstanding tonnage tax regime, introduced some five years ago.

Romania with 0.2 mil. DWT fleet under national flag also use foreign registration (85%). Small fleet of Slovenia even 100% is flagged out. [10]

4 CONCLUSION

We may notice very different approaches to implementation of State aid schemes and maritime policy in European countries. Most of the countries, especially major maritime countries, are using Guidelines as provided in EC Communication C(2004)43 and these Guidelines are to prolonged and improved.

It should be noted also that in a number of European countries there is not enough political will to introduce State aid regime for shipping.

Introduction of tonnage tax system does not necessarily serve as an effective tool for preventing flagging out trends. More complex approach need to be taken. Extension of Guidelines to ship manage-

ment companies is one of the ways which is under consideration in European Commission now.

Other State aid types also are important, such as Marco Polo program and State aid to shipbuilding.

REFERENCES

1. Official Journal of the European Union of 17.1.2004, *C(2004) 43*.
2. Official Journal of the European Union 21.1.2009, *COM(2009) 8 final Brussels.*
3. EC site 15.12.2010 *"Marco Polo - New ways to a green horizon".*
4. Official Journal of the European Union of 30.12.2003, *COM(666).*
5. RS Platou Monthly, December 2010
6. ECSA Annual report 2009 -2010
7. ECSA Annual report 2008 -2009
8. ECSA Annual report 2007 -2008
9. ECSA Annual report 2006 -2007
10. ISL Sipping Statistics Yearbook 2009

13. Sustainable Transport Planning & Development in the EU at the Example of the Polish Coastal Region Pomorskie[1]

A. Przybyłowski

Gdynia Maritime University, Gdynia, Poland

ABSTRACT: The efficient and affordable transport systems are necessary for economic development and for the need to mitigate adverse externalities to health and the environment. Countries all over the world should support greater use of public and non-motorized transport and promote an integrated approach to policy making including policies and planning for land use, infrastructure, public transport systems and goods delivery networks, with a view to providing safe, affordable and efficient transportation, increasing energy efficiency and reducing pollution, congestion and also adverse health effects. The peripheral areas in the EU, especially those situated in regions with the undeveloped accessibility and low level of the economic development, have the opportunity to improve their availability, assuming the proper use of EU resources. The activity of the central, regional and local authorities will be of great importance during the implementation of the adopted development strategies and programmes for transport investments for the period 2007-2013. It is worth taking a closer look at one of the biggest current challenges that is the assurance of sustainable transport development & planning on the regional level. The goal of the paper is to present the research analysis, based on the available strategic documents and statistical data, on the present EU transport policy guidelines in the context of sustainable development concept and cohesion instruments, as well as transport planning and development in Poland with a special regard to one of its coastal regions Pomorskie voievodship.

1 INTRODUCTION

Transportation is expected to be the major driving force behind a growing world demand for energy. It is the largest end-use of energy in developed countries and the fastest growing one in most developing countries. Furthermore, adequate, efficient, and effective transport systems are important for access to markets, employment, education and basic services critical to poverty alleviation. Transport plays an important role in increasing the accessibility of particular regions. Creating development opportunities in peripheral areas through infrastructural investments is one of major EU goals. The peripheral areas in the EU, especially those situated in regions with the undeveloped accessibility - like the coastal region Pomorskie voivodship - and low level of the economic development, have the opportunity to improve their availability, assuming the proper use of EU resources. The activity of the central, regional and local authorities will be of great importance during the implementation of the adopted development strategies and programmes. One of the biggest challenges is the assurance of sustainable transport development planning in compliance with the EU guidelines.

2 EU TRANSPORT POLICY IN THE CONTEXT OF SUSTAINABLE DEVELOPMENT AND COHESION POLICY

In its transport policy the EU aims at changing the demand pattern through shifting potential demand from the road transport sector towards the rail, inland waterway and sea transport – short-distance shipping as well as promoting combined transport and collective public transport. Such solutions are more environmentally friendly, thus helping pursue sustainable development. The transport policy goals are based on two assumptions (Commission of the European Communities, COM (2006):
- mobility is the key to Europe's prosperity and the free movement of its citizens;
- the negative effects of this mobility, i.e. energy consumption and the impact on health and the environment, must be reduced.

The EU transport policy might foster various aspects of the regional development policy pursued

[1] The research paper has been financed out of scientific budgetary resources for the years 2010-2012 as a research project.

within the cohesion policy, and it may influence different sectoral policies implemented by cohesion policy instruments. The functioning of common transport policy instruments brought about many positive EU-wide changes, for instance (Grzelakowski and all., 2008):
– improvement of the quality of services provided and a wider offer of the form and mode of transport,
– reduced costs of transport and a decrease in prices of goods at the Community level, which limited inflation and stimulated exports and investment as well as stabilising the economies of EU Member States,
– improvement of the economic and spatial cohesion of certain parts of the Community,
– improvement of social mobility, resulting in greater labour market flexibility,
– ongoing standardisation of transport equipment and techniques, the development of modern methods and technologies as well as of intelligent traffic management (e.g. interoperability, telematics, the Galileo satellite navigation system).

The EU is fully aware that solely efficient transport sector provided with modern infrastructure and effective market mechanisms can guarantee necessary level of mobility of goods and people. Nowadays, in the age of globalisation and existing highly competitive world economic environment, the mobility is getting essential to the EU's economies and communities. It is key to higher quality of life and welfare as well as fundamental for enhancing EU's competitiveness and vital to achieving the goals of the EU's ambitious strategies for growth and employment.

The mobility, directly connected with the economic expansion (rise of GDP), has been growing in the EU rapidly since the mid of 90s. Goods transport rose ca. 2.8% per year (1995-2006), i.e. more dynamic than GDP did and passenger transport ca. 1.7% per year in the same period. As a result goods and passenger transport grew by 33% and 18% respectively at that time and what is more, this dynamic growth is envisaged to continue in the next decade (see picture 1).

Characteristic trademark of the UE high mobility is, however, relatively outsize share of road transport in the existing modal split. It accounts for 45,6 % in the servicing of total transport demand, whereas rail accounts for 10.5%, inland waterways contribute 3.3% and oil pipelines add another 3.2%. Maritime transport then accounts for 37.3% and air transport for 0.1% of the total traffic (all referring to the EU27 in 2006) (Grzelakowski, 2008).

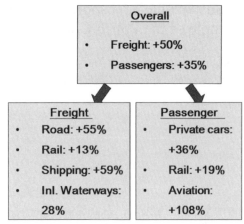

Picture 1. Most likely 2000-2020 growth in transport demand in EU27
Source: Ponthieu E., 'European Economic and Social Committee (EESC). Towards an integrated and coordinated sustainable logistics and transport policy for Europe'. Roma, (19 June 2008), p.10.

As a result of currently formed modal split in the EU's transport sector, and as predicted realistically by 2020, no chance for any shift in it towards the more environmentally friendly modes of transport such as rail and inland waterways, reaching the set up transport policy's objective is thoroughly impossible. When this tendency is followed-up, sustainable mobility by still rapidly growing transport activity will even dash away. For, sustainable mobility this means disconnecting mobility from its many harmful effects for the economy, society and environment (Ponthieu E., 2008). The goals of the EU transport policy stem from the guidelines for development strategies set out at the level of the European Community. The most significant EU strategic documents include the Lisbon Strategy and the Goeteborg Strategy. The former emphasised the necessity to increase the competitiveness of the European area (COM(2005) 24 final), whereas the latter drew attention to ensuring sustainable development of this area (COM(2001)264 final. Recently, the EU has proposed a new document: Europe 2020 Strategy (COM(2010)2020 final). The Commission has identified three key drivers for growth, to be implemented through concrete actions at EU and national levels: smart growth (fostering knowledge, innovation, education and digital society), sustainable growth (making the production more resource efficient while boosting the competitiveness) and inclusive growth (raising participation in the labour market, the acquisition of skills and the fight against poverty).

2.1 Sustainable development concept

As it was defined in the Brundtland Report the sustainable development is development that meets the needs of the present without compromising the ability of future generations to meet their own needs. It contains within it two key concepts (WCED, 1987):
- the concept of **needs**, in particular the essential needs of the world's poor, to which overriding priority should be given; and
- the idea of **limitations** imposed by the state of technology and social organization on the environment's ability to meet present and future needs.

It is possible to graphically represent (picture 2) the achievement of sustainable development by the simultaneous attainment of three objectives: environmental and natural resource sustainability, economic growth and social equity.

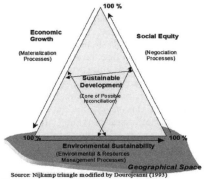

Source: Nijkamp triangle modified by Dourojeanni (1993)

Picture 2. Sustainable Development triangle
Source: Dourojeanni, A. 'Procedimientos de Gestión para el Desarrollo Sustentable: Aplicados a Microrregiones y Cuencas', Santiago: Instituto Latinoamericano y del Caribe de Planificación Económica y Social de las Naciones Unidas (ILPES). Documento 89/05/Rev1., (1993); Nijkamp, P, 'Regional Sustainable Development and Natural Resource Use. In World Bank Annual Conference on Development Economics', Washington D.C., (1990), p.10.

The attainment of environmental sustainability refers to the balance between the human rate of use of the environment and its resources, with natural resources rates of growth and environmental resilience. In similar terms, the attainment of economic growth is related, among other things, to the generation of employment, food, income and wealth (net economic benefits). Social equity refers to the need to give due consideration to the need to generate equal opportunities among people (generational, gender, cultures) to have access to the natural resources base for its use and to the wealth generated. Therefore, the attainment of sustainable development implies the balance between these three objectives or, in other words, to their simultaneous achievement.

Climate change is the most pressing global environmental challenge, and one that calls for major efforts and active steps on the part of industrialised countries, in line with their common and differentiated responsibilities, as well as working in conjunction with transition and developing countries. Any such action must be taken within the framework of the UN Framework Convention on Climate Change (UNFCCC).

According to the Division for Sustainable Development from the United Nations' Department of Economic and Social Affairs, current patterns of transportation development are not sustainable and may compound both environmental and health problems. (www.un.org/esa, 2010). Therefore, there is a need for urgent action, ranging, inter alia, from the promotion of integrated transport policies and plans, the accelerated phase-out of leaded gasoline, the promotion of voluntary guidelines and the development of partnerships at the national level for strengthening transport infrastructure, promoting and supporting the use of non-motorised transport and developing innovative mass transit schemes. The international co-operation is required in order to ensure transport systems support sustainable development. The efficient and affordable transport systems are necessary for poverty alleviation and the need to mitigate adverse externalities to health and the environment. Countries all over the world should support greater use of public and non-motorized transport and promote an integrated approach to policy making including policies and planning for land use, infrastructure, public transport systems and goods delivery networks, with a view to providing safe, affordable and efficient transportation, increasing energy efficiency, reducing pollution, reducing congestion, reducing adverse health effects and limiting urban sprawl (www.un.org/esa, 2010).

There is a need of the full integration of the commitments made by the EU Member States with regard to the Kyoto Protocol and, beyond that, the definition of quantified objectives for reducing greenhouse gas emissions in accordance with the decisions taken by the European Council and the Environment Council in March 2005 - namely to reduce such emissions by between 15 and 30% by 2020 and by between 60 and 80% by 2050, compared with the levels measured in 1990.

2.2 EU cohesion policy

The implementation of the sectoral EU transport policy is supported by the horizontal cohesion policy, especially through structural funds and the Cohesion Fund. The basic goals of the current transport and cohesion policies are shown in Table 1.

Table 1. The golas of the EU transport and cohesion policies.

Goals of the EU transport policy	Goals of the EU cohesion policy
permanent and sustainable development according to the Lisbon and Goeteborg Strategies	sustainable development of all areas preserving the internal economic, social and territorial cohesion through a set of legal and financial instruments
promotion of rail, sea and intermodal transport	solidarity: mitigating the effects of the absence of internal balance at the Community level
integrated regional systems of public transport	cohesion: everyone benefits
development of logistics aimed at obtaining the synergy effect between particular modes of transport and their integration in logistic chains	convergence through investing in infrastructure and human capital, supporting innovation and knowledge-based society, the environmental protection and efficient administration
promotion of intelligent transport systems	regional competitiveness and employment – investing in human resources, entrepreneurship, innovativeness and the development of labour markets fostering social integration
development of trans-European networks	European territorial cooperation – strengthening the cross-border, transnational and interregional cooperation

Source: European Union transport and cohesion policies in the context of rural development, 2008.

It is necessary to support polycentric territorial development of the EU in order to make better use of the available resources in regions (Territorial Agenda, 2007). However, the parameters and monitoring systems to measure territorial cohesion should be defined. Those could be transport accessibility or access to public transport services. Under the transport and cohesion policies attention should be paid to both the territorial cohesion of the whole Europe and the cohesion of specific territories (for example regions), particularly of peripheral areas. It appears that two parallel action strategies might be the solution: the top-down and bottom-up approaches. The former would involve, in accordance with the solidarity principle, the strengthening Community-wide cohesion at the EU level through legal, organisational and financial instruments. The latter strategy would require a regional approach: cohesion development would be initiated by the regions themselves to a larger degree than at present. There is a need for specific financial instruments prepared in agreement with the European Commission to be used, for instance, in the process of creating metropolitan transport systems or cross-border cooperation, as well as in the development of rural infrastructure, especially enhancing access to cities. Such

a system would provide EU support and, at the same time, promote more active regions, mobilising their endogenous potential. It would ensure harmonious development of the whole EU area as well as becoming an important diversifying element. Such a scheme would be competitive, but still stimulating for all the players (Przybyłowski, 2008).

The cohesion policy and its instruments should contribute to the harmonisation of all sectoral policies at the European and national level in order to pursue the Community objectives more efficiently than at present (European Commission, May 2007). But the effectiveness of the EU transport and cohesion policies may be compromised due to significant difficulties as there are some dissimilarities at the implementation level. The transport policy, to a larger degree, aims at liberalisation, free competition, whereas the cohesion policy is more oriented towards interventionism. Therefore, obtaining the synergy effect in regional development and building a coherent and balanced transportation system poses a challenge to the enlarged EU. The key issue is to what extent backward regions should be supported.

It should be emphasised that the development and modernisation of transport infrastructure does not automatically stimulate regional development. While enhancing the economic potential of regions, a comprehensive/integrated approach should be considered so as to ensure that efforts at providing more equal opportunities for the poorest EU areas bring the anticipated results. There are examples of ineffective use of funds throughout Europe, e.g. in East Germany and Greece. Such investment should be coupled with other factors such as material and human capital, the competitive position of local companies, an investment-oriented legal framework (including fiscal regulations), local entrepreneurship. Without those, transport infrastructure cannot become an independent factor of regional development.

As has already been mentioned, the goal of the current EU cohesion policy (see Table 1) is to reduce disparities in the development of particular regions, especially of peripheral areas. This policy is of great significance since it aims at mitigating the effects of the absence of internal balance at the Community level. While creating common policies at the supranational level, the Community remains too concentrated on market processes, neglecting the stimulation of long-term adjustments concerning socio-economic structures. The underlying values can be defined as solidarity and cohesion/harmonisation development. One of them is solidarity since this policy is supposed to be beneficial to citizens and regions in a worse economic and social situation as compared to the EU average,. The other is cohesion because everyone would benefit from reduced disproportions in income and well-being between the poorer and wealthier countries and regions. The degree of such disparities is measured in three aspects:

economic (mainly by the purchasing-power-parity-based GDP per inhabitant of the region), social (*inter alia* by the unemployment rate in the region) and spatial (usually by a measure of the number of consumers over a given period in a given region) (http://ec.europa.eu/regional_policy, 2011). Structural indicators are also important. They are used by the European Commission in the evaluation of the EU Member States' progress in the implementation of the Lisbon Strategy goals. They include five main socio-economic domains of employment, innovation and research, economic reform, social cohesion and the environment, as well as the general economic background.

In 2007, the EU introduced a modernised and more integrated cohesion policy. It covers the period between 2007 and 2013. The combined budget of structural funds and of the Cohesion Fund in this period will amount to ca. EUR 308 billion, accounting for 36% of the total EU expenditure in the period in question. Three funds are the instruments of the amended cohesion policy: the European Regional Development Fund (ERDF), the European Social Fund (ESF) and the Cohesion Fund. The appropriations were divided into three categories. 81.5% of the total amount was assigned to reducing the disproportions between the poor and wealthy regions (the Convergence objective), while 16% – to the improvement of the competitiveness of the poor regions and job creation (the Regional competitiveness and employment objective). The remaining 2.5% is aimed at supporting cross-border cooperation between frontier regions (the European territorial cooperation objective). It should be emphasised that the compensatory nature of the cohesion policy (in response to the needs of lagging regions) in the amended Lisbon Strategy of 2005 was replaced with active creation of conditions for development. At present, the focus is on the promotion of competitiveness and creating new jobs, not only on standard convergence activities. Thereby the gap between the EU pursuit to increase its competitiveness on the one hand, and to support regions merely to reduce differences on the other hand is diminishing.

Authors of some analyses point out that the concentration on connecting regional capitals in new Member States may contribute to increasing the differences within these countries and lead to an anticohesion effect. Due to the focus on the development of TEN-T networks, the EU actually marginalises expenditure on the remaining transport networks, which leads to the imbalance between European and regional projects. Cohesion reports unambiguously show that as the cohesion between Member States grows, the development gap between particular regions within these countries widens. Unfortunately, this negative trend is also observed in Poland.

3 TRANSPORT NETWORK PLANNING AND DEVELOPMENT IN POLAND

The transport system in Poland is neither sustainable nor efficient in economic or technical terms, which entails specific environmental and social consequences. From the point of view of Poland's transport needs, the accession to the European Union in 2004 created new possibilities in the field of extension and modernisation of transport infrastructure since within the framework of the common transport policy and cohesion policy there are instruments and funds available for these purposes. At the same time, Poland's membership in the European Union involves the introduction of and compliance with a number of requirements concerning transport infrastructure.

The present condition of transport infrastructure in Poland does not meet the expectations of users of national roads, railways and other transport sectors. It also fails to provide appropriate handling of international cargo flows under the rapid growth in traffic, which has been observed for more than a decade. Furthermore, transport users have been increasing their requirements regarding the quality of transport services, in particular reduced transport time, improved safety and ensuring intermodality of the transport process. Significant decapitalisation of infrastructure facilities and equipment as well as not always appropriate spatial distribution of specific network elements may maintain or generate regional disproportions within Poland. Major infrastructural gaps can be found in all the transport sectors. Due to the absence of an appropriate network of motorways, express roads and high-speed rail system, the existing transport network structure does not contribute to the effective allocation of resources and does not ensure appropriate quality of passenger and cargo transport. Sea ports, inland waterway ports and airports should also be modernised.

The most important tasks in the field of road infrastructure development from 2007 to 2013 include:
– extending the network of motorways and express roads;
– programme of improving the pavement on roads where heavy truck traffic can be observed;
– eliminating the shortcomings in the current road network maintenance;
– programme of building by-passes or ring roads around towns, ensuring that such roads are secured against new building developments;
– modernisation of national road sections aiming mainly at improving traffic safety, including the launch of a programme for reducing traffic on roads running through small towns and villages,
– improving the conditions for transit traffic as well as for origin-destination traffic within metropolitan areas.

The special Operational Programme: Development of Eastern Poland comprises plans to build or modernise road sections which will contribute to improving connections between the most peripheral parts of Poland and the transport network.

The density of gminas (the basic unit of the country's territorial structure) roads in Poland was 47.8 km per 100 km², while the overall length of gminas roads amounted to ca. 150,000 km at the end of 2004. At the same time, the density of access roads to agricultural and forest land was 90.1 km per 100 km² and their overall length reached ca. 289,000 km. Spatial distribution of roads is strongly connected with population density and economic characteristics of the area in question, therefore the highest density of the road network is found in the Małopolskie, Śląskie, Opolskie, Dolnośląskie and Wielkopolskie voivodships. The rather well-developed network of access roads to agricultural and forest land is nevertheless characterised by very low pavement quality. At the same time, the quality of gminas roads is directly connected with bus communication networks (both municipal and private), which enable local residents to get to urban centres and to commute to their non-agricultural jobs. It is of great importance particularly in the context of the liquidation (due to low profitability and financial inefficiency of local governments) of regional rail connections in many voivodships. Thanks to EU support it will be possible to reduce this development gap.

However, as regards the development of local roads, one of the reports carried out for the Ministry of Regional Development indicates that such roads do not form a coherent network and are not sufficiently integrated into the voivodship development strategy implementation. Considering IROP projects implemented so far, the complementarity index for local roads (ranging from 0 to 3) was 1.6 on average. The Podkarpackie, Świętokrzyskie and Lubelskie voivodships used the EU support the most efficiently, whereas the worst performer was the Pomorskie voivodship (http://mrr.gov.pl, 2011-02-01).

There is a need for instruments increasing the innovativeness of technical solutions in the field of transport infrastructure and therefore providing a greater choice between various modes of transport. The routine approach to increase the number of roads and motorways, consisting in allocating most funds to these goals, contradicts the principle of sustainable development. After decades of intensive development of road infrastructure in the EU-15, for ca. 20 years a greater emphasis has been put on the improvement of the railway, inland and sea transport infrastructure. Similar observations can be made as regards the improvement of public transport systems in major European cities, used by a growing number of commuters who switch from passenger cars to public transport. Integrated regional public transport systems represent an EU requirement: Poland is obliged to implement this directive by 2013. The integrated regional public transport systems include integrated tickets covering all means of public transport, along with numerous systems of group, zone or time discounts encouraging passengers to choose public transport services. Such systems are also strengthened by the policy of imposing very high parking charges in the cities, or by locating parking lots for bicycles near train or underground stations. Such solutions are yet to be introduced in Poland. The maturity of urban communities and switching to integrated urban transport services will become a new qualitative factor affecting the structure of demand for transport (Burnewicz, 2008).

Finally, there is a need to combine the processes of extending necessary transport infrastructure with the rule of balancing development by seeking selective and optimal solutions at the level of regions and at the local level. Other instruments include much wider application of the principle of genuine rather than only facade social participation in the decision-making on roads, motorways and other infrastructural lines, in order to balance the interests of local and regional communities and their development ambitions as well as taking account of environmental protection aspects in investment processes in a much more strategic way than it was the case in the past (Gończ, 2007). In Poland, further decentralisation of the state and public finance, along with a more extensive scope of decisions taken at the regional level would also contribute to the harmonisation of investment activities and sustainable development challenges.

4 TRANSPORT INVESTMENTS PLANNING AND DEVELOPMENT ANALYSIS IN THE POMORSKIE COASTAL REGION

Considering the social and economic situation as well as the SWOT analysis for the voivodship, the authorities of the Pomorski region prepared the Development Strategy for the Pomorskie voivodship until 2020 (www.woj-pomorskie.pl, 2007); the strategy aims at overcoming the weaknesses in order to make the best possible use of the opportunities.

It is compliant with the strategic goal covered by the NSRF[2], envisaging the Pomorskie Voivodship of 2020 to be an important partner in the Baltic Sea region,– offering a clean environment, high quality of life, development driven by knowledge, skills, active and open communities, a strong and diversified economy, cooperation based on partnership, an attractive and coherent area, conserving multicultural heritage as well as solidarity and maritime traditions.

---

[2] The goal under the NSRF is the creation of the conditions for improving the competitiveness of knowledge-based economy and entrepreneurship ensuring an increase in employment and greater social, economic and territorial cohesion.

The implementation of this vision is based on three new priorities, strategic objectives and specific courses of action (Table 2).

The voivodship authorities were obliged to develop a Regional Operational Programme for the Pomorskie Voivodship for 2007-2013 as an instrument for the implementation of the NSRF within the region and, at the same time, a document enabling EU support to be obtained under the Community regional policy objective "Convergence." The programme is in line with the provisions of the following (ROP, 2007):
- - Development Strategy for the Pomorskie Voivodship,
- - National Strategic Reference Framework,
- - Community Strategic Guidelines on Cohesion.

Table 2. Priorities and strategic objectives of the Pomorskie voivodship until 2020.

COMPETITIVENESS	COHESION	ACCESSIBILITY
1. Improved conditions for enterprise and innovation	1. Employment growth and increased labour mobility	1. Efficient and safe transport system
2. High level of education and research	2. Strong, healthy and integrated society	2. Improved operation of technical and ICT infrastructure systems
3. Development of an economy based on specific regional resources	3. Civil society development	3. Better access to social infrastructure, particularly in structurally disadvantaged areas
4. Efficient public sector	4. Shaping social and spatial processes to improve the quality of life	4. Conservation and improvement of the natural environment
5. Established position and effective links between the Tri-City Metropolitan Area (*Trójmiasto*) and other, mainly Baltic, regions	5. Strengthening sub-regional development centres	

Source: Development Strategy for the Pomorskie Voivodship–July 2005), www.woj-pomorskie.pl/downloads/ASRWP_tekst, 2007-08-09, p. 23.

The overall strategic objective of the Programme is therefore the improvement of economic competitiveness, social cohesion and spatial accessibility through sustainable use of specific features of the potential. ROP financial instruments using the EU structural funds are shown in Table 3.

As shown in Table 3, the voivodship authorities intend to allocate the highest share of the funds (23%) for the development of the regional transport system, which may be regarded as a good decision since the transport system in the Pomorskie voivodship is inefficient. Major shares of the appropriations will also be granted to small and medium-sized en-

terprises (21%), basic local infrastructure (14%) and projects concerning the development of metropolitan functions (12%). A relatively small amount has been provided for tourism and cultural heritage (only 5%); the lowest share of funds was allocated for technical assistance (3%). The regional transport system (priority axis 4) in the Pomorskie voivodship will receive a total of EUR 271,420,167 (with the Community contribution of 75%). As regards other priority axes of importance to infrastructure development, the following are worth mentioning : axis 3 concerning urban and metropolitan functions (over EUR 150 million), axis 6 regarding tourism (almost EUR 60 million) and axis 8 aiming at the improvement of basic local infrastructure (more than EUR 145 million). A strong preference will be given to projects in line with the development programmes of the whole transport infrastructure system covering all sectors and following from the Transport Development Strategy of the Pomorskie voivodship.

Table 3. The structure of ERDF funds allocation by Priority Axis of ROP PV.

Priority axis	ERDF funds allocation (%)
1. Development and innovation in SMEs	21.0%
2. Knowledge-based society	7.0%
3. Metropolitan functions	12.0%
4. Regional transport system	**23.0%**
5. Environment and environment-friendly power industry	7.0%
6. Tourism and cultural heritage	5.0%
7. Healthcare and rescue system	4.0%
8. Basic local infrastructure	14.0%
9. Local social infrastructure and civil initiatives	4.0%
10. Technical assistance	3.0%
Total	100.0%

Source: Own study based on: ROP (Regional Operational Programme) for the Pomorskie Voivodship 2007-2013, Annex to Resolution of the Pomorskie Voivodship Executive Board No. 75/18/07), 5.02.2007, p. 64.

ROP PV will be financed from the ERDF as well as with national funds, and the contribution from the ERDF – according to Council Regulation No. 1083/2006 – was calculated with reference to the total eligible expenditure, including public and private expenditure. The amount allocated to investment will total EUR 1,227.1 million, of which the national public and private contribution will be EUR 240.7 million and EUR 101.4 million respectively. Almost half of the budget will be used for the implementation of the Lisbon goals. Other funds from other programmes under the EU cohesion policy, the common agricultural policy and national policies and strategies will also be of considerable importance (Table 4). (www.mrr.gov.pl, 2008).

The competitiveness and cohesion of each region largely determine the condition and development prospects of transport infrastructure. The transport system of the Pomorskie voivodship consists of all types of land, water and air transport (picture 3).

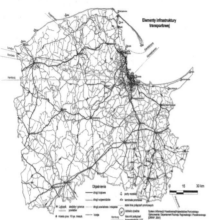

Picture 3. Transport infrastructure of the Pomorskie Voivodship coastal region.
Source: Development Strategy…, op. cit., www.woj-pomorskie.pl/downloads/ASRWP_tekst, 2007-08-09, p. 15.

The main development problem of the region is the low quality and limited coherence of the transport system. Despite the good location at the crossing of two transport corridors, transport accessibility of the voivodship is quite low against other central and southern regions of Poland and the EU. Western and eastern parts of the voivodship require the improvement of accessibility and quality of transport connections with the regional economic centres, mainly with the Tri-City agglomeration (*Trójmiasto*). The road network does not ensure good access to Gdynia and Gdansk ports. Low quality of transport infrastructure prevents appropriate quality of passenger and cargo transportation services. The current condition increases business costs, lowers the efficiency and competitiveness of companies, thus reducing the attractiveness of the region for foreign investors. It also has a negative impact on the residents' quality of life.

The road network of the voivodship is over 19,500 km long and covers: 8 national roads, 69 voivodship roads as well as *poviat* and *gminas* roads. There are almost no roads of the highest technical standard, and the majority of roads in the region are of low quality and require modernisation. Another weakness is the poor technical condition of bridges and overpasses, of associated infrastructure and of infrastructure related to traffic safety and organisation. Moreover, a significant development barrier is the insufficient capacity of some road sections and the absence of ring roads for transit traffic. Due to reduced cargo and passenger traffic, the overall length of the railway network is also gradually decreasing. The railway lines currently in use in the Pomorskie voivodship are limited to 1,308 km (den-

sity of 7.2 km/100 km^2). The following railway lines included in the Trans-European Transport Network (TEN-T) run through the voivodship: line E-65 (Gdynia-Warszawa-Zebrzydowice), CE-65 (Katowice-Tczew) and Gdynia-Kaliningrad line. As compared to other transport modes, rail transport fails to be competitive. Railway lines and the rolling stock suffer quick decapitalisation, and more and more regional lines are being closed.

In recent years air transport in the Pomorskie voivodship has been characterised by a rapid growth in traffic. The Lech Walesa Airport in Gdansk plays a dominant role in the handling of passengers. For example, in 1991-2005 the volume of cargo doubled, and the number of passengers carried increased almost eight times. Forecasts of increased air traffic point to the need of extending the airport and putting other airports in the voivodship into operation, not as yet used by civil aviation, to serve as complementary facilities. The voivodship authorities decided on situating such an airport in Gdynia – Babie Doły.

Finally, it should be mentioned that mere investment in transport is not enough to stimulate economic growth in the regions. There is a need for rational strategies and regional programmes to include infrastructure investments in a wider context (Parteka, 2007).

The support for regional development via EU instruments brings about improved territorial cohesion of some areas, like Pomorskie region. At the same time, there are also negative results of allocating the European funds for the implementation of the objectives set out by these policies, especially as regards peripheral areas, which leads to neglecting certain aspects, e.g. transport connections between metropolitan areas, towns and villages.

5 CONCLUSIONS

1 Sustainable transport planning and development is a great challenge for the EU, national and regional authorities. Neglecting the development of regional and local transport networks (e.g. via the extension of trans-European networks) can be an example of such a dilemma. Another problem is excessive concentration of expenditure on infrastructural objectives which are not properly linked to other development measures or, for instance, at the expense of innovation measures.

2 Two main dimensions of the EU transport policy, i.e. reduced environmental pressures and sustainable mobility of human resources are significant for other EU policies, e.g. with regard to improved transport in cities and metropolitan areas or support for the development of polycentric networks.

3 Despite the declared willingness to pursue sustainable development at the level of operational

documents drawn-up by the government administration, in Poland the most funds are allocated to road infrastructure (national roads: 33.3%, motorways: 16.6%). This is also the case in the Pomorskie voivodship, although environment-friendly projects are given more attention due to the coastal location of the region. However, the co-financing rate for infrastructure projects still represents a significant obstacle. The EU contribution of up to 75% (and in the case of some investments only 50%) may pose a major problem to many potential beneficiaries within the region.

4 The case of Pomorskie coastal region proves that it is necessary to diversify transport investments in order to ensure sustainable development, which could be fostered, *inter alia*, by integrated regional public transport systems. Partnership based on an extended and efficient institutional cooperation network, coordinated by voivodship governments and covering local and regional authorities, socio-economic partners, universities, business organisations, non-governmental organisations, government institutions, as well as other Polish and foreign regions and institutions, might also prove helpful in the sustainable transport planning and development implementation.

REFERENCES

Burnewicz J., Wizja struktury transportu oraz rozwoju sieci transportowych do roku 2033 ze szczególnym uwzględnieniem docelowej struktury modelowej transportu, http://www.mrr.gov.pl/NR/rdonlyres/, 2008-01-15, p. 5.

COMMUNICATION FROM THE COMMISSION EUROPE 2020: A strategy for smart, sustainable and inclusive growth Brussels, 3.3.2010, COM(2010) 2020 final.

Communication from the Commission to the Council and the European Parliament, "Keep Europe moving – Sustainable mobility for our continent. Mid-term review of the European Commission's 2001 Transport White Paper", Commission of the European Communities, COM (2006) 314 final, Brussels 2006.

Communication from the Commission to the European Parliament, the Council, the European Economic and Social Committee and the Committee of the Regions, 'Strategic goals and recommendations for the EU's maritime transport policy until 2018', COM/2009/0008 final.

Communication from the Commission, 'A sustainable Europe for a Better World: A European Union Strategy for Sustainable Development', Brussels 15.05.2001, COM(2001)264 final.

Development Strategy for the Pomorskie Voivodeship– July 2005, www.woj-pomorskie.pl/downloads/ASRWP_tekst, 2007-08-09.

Dourojeanni, A. 'Procedimientos de Gestión para el Desarrollo Sustentable: Aplicados a Microrregiones y Cuencas', Santiago: Instituto Latinoamericano y del Caribe de Planificación Económica y Social de las Naciones Unidas (ILPES). Documento 89/05/Rev1., (1993).

EU funds for the Pomorskie Voivodship in 2007-2015, www.mrr.gov.pl, 2008-10-31.

European Commission, 'ENERGY AND TRANSPORT IN FIGURES 2007', Directorate-General for Energy and Transport in co-operation with Eurostat, p. 8.

Gończ E., Ulf Skirke, Hermanes Kleinzen, Marcus Barber, *Increasing the Rate of Sustainable Change: A Call for a Redefinition of the Concept and the Model for its Implementation*, ELSEVIER, Science Direct, Journal of Cleaner Production 15 (2007)

Growing Regions, growing Europe. Fourth report on economic and social cohesion, Communication from the European Commission, May 2007.

Grzelakowski A. S. G., "European greener mobility", Baltic Transport Journal, (2008).

Grzelakowski A. S., Matczak M., Przybyłowski A., *Polityka transportowa Unii Europejskiej i jej implikacje dla systemów transportowych krajów członkowskich*, Publ. AM in Gdynia, Gdynia 2008 (in press), p. 66.

http://www.un.org/esa/dsd/susdevtopics/sdt_transport.shtml, (2010-07-10).

http://mrr.gov.pl, 2011-02-01.

Networks for peace and development. Extension of the major trans-European transport axes to the neighbouring countries and regions, November 2005, www.eu.int/comm, (2007-02-23).

Nijkamp, P, 'Regional Sustainable Development and Natural Resource Use. In World Bank Annual Conference on Development Economics', Washington D.C., (1990).

Parteka T., Przemysły morskie i infrastruktura techniczna w Strategii Rozwoju Województwa Pomorskiego do 2020 roku, (in:) A. S. Grzelakowski, K. Krośnicka (eds.), Przemysły morskie w polityce regionalnej UE, Gdynia Maritime University, Gdynia 2007

Ponthieu E., 'European Economic and Social Committee (EESC). Towards an integrated and coordinated sustainable logistics and transport policy for Europe'. Roma, (19 June 2008), p.10.

Przybylowski A., European Union transport and cohesion policies in the context of rural development, Warsaw 2008.

Przybyłowski A., Zintegrowane podejście do polityki rozwoju Unii Europejskiej – polityka spójności a polityka transportowa, (in:) Grosse T., Galek A. (eds.), Zintegrowane podejście do rozwoju. Rola polityki spójności, Ministry of Regional Development, Warsaw, June 2008, pp. 119-158.

ROP (Regional Operational Programme) for the Pomorskie Voivodship 2007-2013, Annex to Resolution of the Pomorskie Voivodship Executive Board No. 75/18/07), 5.02.2007

Territorial Agenda of the European Union. Towards a More Competitive and Sustainable Europe of Diverse Regions, Leipzig 2007.

The EU regional policy – overview, http://ec.europa.eu/regional_policy, 13-01-2011.

World Commission on Environment and Development (WCED), 'Our common future', Oxford: Oxford University Press, (1987), p. 43.

14. Development of the Latvian Maritime Policy; A Maritime Cluster Approach

R. Gailitis
Latvian Maritime Administration, Riga, Latvia

M. Jansen
Netherlands Maritime University, Rotterdam, Netherlands

ABSTRACT: Latvia is a maritime nation although most of its inhabitants are unaware of the importance of maritime activities to the economy. For policy makers and also for stakeholders of the maritime clusters it is important to understand which factors contribute for the sustainable development of the companies in the cluster. Therefore the aim of this paper is to analyse the economical value, which gives the possibility to assess the importance of maritime resources for country. On the basis of this analysis, the authors conclude an integrated approach should be applied which is based on knowledge about the economical importance of the maritime sectors, their economical links and their strategic tradeoffs for future development.

1 INTRODUCTION

Latvia is a maritime nation due the local maritime resources like ports, seafarers etc. However, there is a lack of awareness in the community about development prospects of Latvia's maritime sector. Does Latvia need maritime resources? This is a frequently asked question made by in the public opinion. There seems to be a lack of information about maritime resources and their contribution to country's economy. There is no updated study on the importance of Latvian maritime resources and their contribution for Latvian economy at present and how valuable for Latvian economy they can be in the future.

Looking to present situation, the current maritime knowledge infrastructure is not supported by a deliberate policy to strengthen the stakeholders within this sector of the Latvian economy. It is merely based on legal principles, most of them to accommodate the standards set by the International Maritime Organisation (e.g. STCW) and translate to country specific rules and regulations. Having an effective policy requires much more as the policy is a way to influence business, living and social climate with the aim to push the economy in the right direction. Up until now, the Latvian government has been steerless in this respect. If governments do not know what is the game played on the high seas, it is difficult to manoeuvre your own little boat through safe havens...

This article will explore the economic value of seafarers for the Latvian economy thereby striving to create awareness and initiate an action plan to develop a coherent maritime policy to support sustainable development of the maritime cluster and companies within.

2 LATVIAN MARITIME CLUSTER IN CONTEXT OF M. PORTER'S CLUSTER THEORY

2.1 *The cluster concept*

The cluster concept tries to put into the frame a business environment and considers the possibilities of the development. Therefore it can be concluded that cluster itself is an environment. Regarding the Porters definition for a cluster it does not put strict margins on a cluster as such and it is based on empirical studies from different industries.

M. Porter (1998) stated that: "Clusters are geographical concentrations of interconnected companies, specialised suppliers, service providers, firms in related industries and associated institutions (for example, universities, standards agencies, and trade associations) in particular field that compete but also cooperate." Actually there still is going an on-going debate on what constitutes a cluster, both among academics and among policymakers, and there are a multiple perceptions of the kinds or the categories of the clusters (Andersson, Schwaag Serger, Sörvik, Hansson 2004). However, as most important factors for determination and investigation of cluster can be mentioned geographical concentration, critical mass of companies, multiple actors and active business channels between stakeholders which involve cooperation and competition.

2.2 Cluster concept in perspective of the Latvian maritime cluster

The Latvian maritime cluster is not defined in any official policy document. With regard to the presence of the maritime cluster it can be seen that the building stones for the maritime cluster within the business environment of Latvia exist. There are numbers of companies active in different maritime business sectors located in Latvia. The European Cluster observatory distinguished a number of sectors, which together make up the maritime cluster. All of these maritime sectors are represented in Latvia.

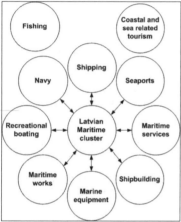

Figure 1: Latvian maritime cluster according to the methodology of European cluster observatory

However, the links between those maritime sectors are weak and the maritime cluster is deemed to be underdeveloped. Some of the sectors are more developed than others. Shipping, seaports, shipbuilding and maritime services can be considered as being the most observable sectors in the Latvian maritime cluster.

3 MAIN MARITIME SECTORS IN LATVIA AT A GLANCE

3.1 Shipbuilding sector

Shipbuilding in Latvia is represented by four enterprises predominantly active in the ship repair sector (additionally there are some small enterprises, largely associated with in house servicing of fishing or small river vessels). As the shipbuilding is active in ship repair activities there are weak economical links between shipping and shipbuilding sectors as the Latvian ship owners are not purchasing their ships at Latvian shipyards.

3.2 The port sector

There are 3 big and 7 small ports. Approximately 70% of all cargoes going through Latvian ports are transit cargoes. Due to the cargo flows through Baltic area and related business which directly focuses on those cargo flows, like road transport, rail transport, shipping agencies, ferry services from Latvian ports, freight forwarding etc. it can be assumed that the ports' role in logistic cluster is important. However, as the logistic cluster is almost independent of the rest of the maritime cluster it is difficult to define the real value of the ports in the logistic cluster and induced effects on rest of the maritime cluster.

3.3 Merchant shipping

The merchant shipping sector can be characterized with a number of small shipping companies. According the data from Equasis and Sea Web (Lloyd List data base) there are located around 30 shipping companies which operate fleet of about 150 ships with total GT 1,408,243. However, only about 40 ships are owned by national ship owners. The rest of the ships are owned by foreign owners, which has located their shipping companies in Latvia. Latvian Shipping Company which owns around 21 ships can be considered as the main player. With regard to the available data the size of national fleet during last 18 years has considerably decreased (see table 1).

Table 3 Number of ships owned by national ship owners

Year	1992	2000	2010
Number of owned ships by Latvian shipowners*	110	80	40
Number of owned ships by LSC	89	60	21

* Estimated number of ships considering the number of companies owned by national owners

The national seafarers can be considered as the other part of merchant shipping sector. There are around 13,000 seafarers of whom around 12,000 are active at the merchant fleet. However the number of the seafarers during last 6 years has decreased (see table 2).

Table 2 Changes of the number of the number of Latvian seafarers working on merchant vessels (2005-2011)

	01.01.2005	01.01.2011	Difference
Ratings	8500	6616	-22%
Officers	5734	5691	-1%
Total	14234	12307	-14%
Deck officers	2383	2541	7%
Engine officer	2541	2625	3%
Non conventional officers	679	552	-19%

The main decrease has been on account on the number of ratings. According the data from Latvian Seamen Registry the decrease of ratings has been 22% in comparison with number of ratings in 2005. The decrease of ratings is linked with increased competition with ratings from Asia and economical growth of Latvian economy during (2004 -2007) when part of rating switched to work ashore in construction companies. The decrease of officers during this period is not so considerable and is linked with decrease of number of non conventional engine officers like electrical engineers and gas engineers. The decrease of non conventional engine officers is linked with lack of maritime education programs for them. However recently the maritime education program for electrical engineers has been re-established and the first graduates joined pool in 2010. The impact of this programme on number of non conventional engineers will have effect in closest year. The positive aspect is that number of conventional officers both deck and engine has been increased during this period.

3.4 Maritime services

There is a wide range of maritime service businesses in Latvia. The maritime services can be separated in to the two groups – services which focus on shipping sector with core subject of business "ship" like shipbroking, maritime law, maritime insurance, and services which focus on shipping sector with core subject of business "seafarers" like maritime education and training, crewing agencies etc. The first group of services with focus on ship is not developed as the shipping sector itself in Latvia is not developed.

Due to the weak home demand from Latvian shipping companies for those services the services are not internationally competitive.

The other part of the maritime services with focus on "seafarers" can be considered as well developed. All services which seafarer would need are available. As an example there are established around 50 crewing agencies and 8 training centres, which are considerable numbers for such a small country as Latvia is. The crewing agencies not only provide recruit Latvian seafarers but also recruit seafarers from Belorussia, Ukraine, Russia therefore adding value for Latvian maritime cluster.

4   ECONOMICAL VALUE OF SEAFARERS

Economical value gives possibility to assess the importance of maritime resource for the country.

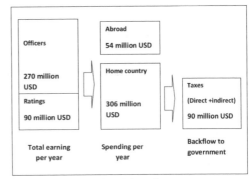

Figure 2: Economical value of the seafarers

According the authors' analysis, the pools of seafarer positively contribute to Gross National product of Latvia and other economical processes in country due to money which they spend in Latvian economy. The total contribution to Gross National product is around 1.3 percent in 2010. Comparing with average employee in Latvia ship's officer earn around 4.5 times more than average employee in Latvia while average rating earn only 1.3 time more than average employee. This lead to conclusion that officers are much more valuable for Latvian economy, even they are less in number than ratings.

Regarding the considerable number of the seafarer resources it can be seen that the subcluster is formatted around them. The seafarers cluster included crewing companies, maritime education centres, nongovernmental associations, training centres and other stakeholders who benefit from presence of large pool of seafarers.

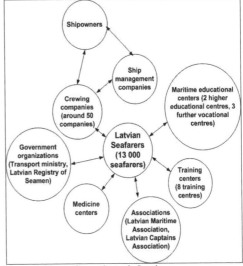

Figure 3: The seafarers cluster in Latvia

Considering the present economical value of the national seafarers it can be concluded that they are an important driver for Latvian economy especially during the time of economical recession of national economy as they are bringing money from foreign shipowners and spend their earnings in Latvian economy.

5 LINKS BETWEEN MARITIME SECTORS IN LATVIA

Demand and supply links, the so-called factor conditions according the Porter's Diamond model inter-link the maritime sectors within a cluster. So for the maritime cluster it is important to have those demands/supply links between the players of the maritime cluster as the growth in one sector will induce the growth in other sectors as well. Considering the general maritime cluster, the shipping and related industries like (navy, cruise shipping, fishing) are main demand generators in the rest of maritime sectors. However, nowadays shipping are looseing its links with the other maritime sectors on their home market due to the globalisation. Shipowners operate their ships far away from their home country and are free to order their services in any country as long as it is cheap and convenient.

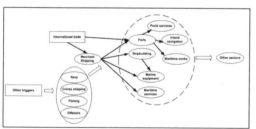

Figure 4: Demand supply relationships in maritime cluster (general model)

Also the Latvian maritime cluster is not an exception to this rule of thumb, considering the most observable sectors it can be seen that ports depend on the local transit cargo flows, but the shipping sector is not attached to these cargo flows. Therefore it leads to the situation that the two most important sectors in Latvian maritime cluster have different interests and weak links. Due to lack of critical mass in shipping sector the links between ports and shipping sector is limited. The port sector and related sectors are almost independent from shipping sector and other maritime sectors. The shipbuilding also is not linked with shipping sector as the shipbuilding focuses on ship repair activities and Latvian shipowners cannot order their ships through Latvian shipyards.

6 CONCLUSIONS

Latvia is a maritime nation as it can be derived from the local maritime resources like ports, seafarers and related business activities. However, there is a lack of awareness in the community about development prospects of Latvia's maritime sector. This article provides an updated overview of the maritime sectors and placed these economical activities in a conceptual framework based on the Porter's cluster theory. Most important elements of a well-developed cluster are: geographical concentration, critical mass of companies, multiple actors and active business channels between stakeholders who are in cooperation and competition with each other. This article explores the economical value of Latvian seafarers in order to assess their contribution to the Latvian economy, thereby striving to provide awareness and sense of urgency for future development of the maritime cluster in Latvia.

Latvian maritime cluster is lacking some key elements like critical mass in shipping sector, shipbuilding sector and maritime equipment sector, therefore the links and cooperation between those sectors are weak.

One of the problems for the maritime cluster development is lack of the successfully performing maritime policy. This factor impedes the process of the development of maritime cluster. There also is no shared vision, mission, goals and strategy among the stakeholders of Latvian maritime cluster. The cluster mindset is not developed in maritime community and therefore it is important to understand what is cluster and which factors can contribute for successful cluster development.

In line with Porter's cluster theory, there is a task of the government to set a policy which aims to create an environment where companies can build critical mass to create inner dynamics for the maritime cluster development. This would increase the economical value of cluster and increase the backflow for government. The other aim for strengthening maritime cluster is to establish policy measures which would help to interlink them in a better way. As the policy as paper is living document it should be subject of periodical review to diminish risk for policy to become outdated and for nation to lag behind other nations.

At the first stage the maritime policy should be written down explicitly. As the maritime cluster is a complex business environment, an integrated approach should be applied based on knowledge about the economical importance of the maritime sectors and their economical links.

6.1 Factors which should be integrated in maritime policy

6.1.1 Focus strategy

The focus strategy of the Latvian maritime cluster probably can be Short sea shipping in Baltic/Nordic region or transit cargoes towards Russia, but this does not imply that other areas in maritime cluster do not get attention in the new policy. Important activity of Latvian maritime cluster is supply of seafarers for EU shipowners therefore Latvian maritime cluster can be linked to EU market as a new broader home market which would help to develop home demand conditions (according model of Porter) in the Latvian maritime cluster;

6.1.2 Stimulation

The maritime policy should focus on the stimulation of the "home demand condition" which would influence the factor conditions. The stimulation of the "home demand" is essential as the main problem of the Latvia maritime cluster is lack of critical mass of the companies. The home demand conditions can be stimulated by making available capital through a finance system like KG ship finance structure in Germany and by linking home demand to other EU countries, which requires the cooperation between countries at international level.

6.1.3 Attractiveness of the Latvian maritime cluster

It is recommended to develop Latvian maritime resources to stimulate local maritime business and attract foreign users to use those resources through benefits for companies and seafarers stimulating economic activity of the maritime cluster.

6.1.4 Communication and cooperation

It is recommended to set up maritime cluster organisation which would provide an environment for communication and cooperation between policymakers and the rest of the stakeholders.

6.1.5 Support for innovation and education

Although the role of innovations has not been topic of this research, is obvious that support for innovations and maritime education is essential. The innovations can help in creation of the special factor conditions which would serve as unique competitive advantages. Therefore it is recommended to set up innovation policy to support creation of the research and development centre. The support for maritime education is essential as the maritime knowledge is inherited and embedded in the maritime education of the country. The available maritime knowledge can provide maritime cluster with the unique competitive advantages and serve for sustainable development.

6.2 Further research

Further research is required to gain more knowledge and information about the economical importance of the different maritime sectors. Due to the limitations of this article, only the economical importance of Latvian seafarers has been assessed. The same quantification needs to be done for other maritime sectors.

A second part of such research should focus on the economical value of the links between the maritime sectors, i.e. what is the purchase value of one sector to the other and what is the import and export value of maritime services from Latvian based maritime companies. This should help to comprehend how the development policy of maritime cluster can be set to achieve the proposed aims.

Based on the economical value of the maritime cluster as a whole and the understanding both the linkages, governments should work out its policy instruments in line and in close cooperation with business strategies of companies acting in the cluster. Authors put strong emphasis on the sustainability of the maritime policy. The effects of these policy instruments should be subject to a longitudinal monitoring survey, measuring the economical value, supply and demand on maritime employment and education and innovation in the forthcoming years.

REFERENCES

Gailitis, R. 2010. *Development of the Latvian maritime policy; Maritime cluster approach, Master thesis*. Rotterdam: Netherlands Maritime University

Andersson, T. & Schwaag Serger, S. & Sörvik, J. & Hansson, E.W. 2004. *Cluster Policies White Book; Proc. of the 6th Global Conference of The Competitiveness Institute (TCI), Gothenburg, Sweden, 17-19 September 2003*. Malmo: International Organisation for Knowledge Economy and Enterprise Development

Porter, M.E. 1998. *On Competition*. Boston: Harvard Business School Press

Maritime Law

15. European Union's Stance on the Rotterdam Rules

L. Kirval
Maritime Faculty, Istanbul Technical University, Turkey

ABSTRACT: In today's rapidly globalizing world economy, the importance of maritime transportation is increasing. Today, approximately 80% of the global transportation services is done by the seas. Therefore, the international laws and regulations that outlines the rights and responsibilities/obligations of the carriers and cargo owners is of very high importance for the smooth running of this global maritime transportation system. On the other hand, today, "door to door" and "multimodal" transport is getting widely used. However, during this type of highly complicated transportation, the rules and the applicable laws with regards to rights and obligations of the parties (carriers and cargo owners) greatly vary, and this creates several problems particularly about the carrier's liabilities.

For offering solutions to these problems and creating world-wide uniformity about the carrier's and cargo owners' rights and obligations, United Nations Commission on International Trade Law (UNCITRAL) has offered alterations to the current international regulations in force (which are now generally called as Rotterdam rules), and, to this date, more than 20 countries have signed this new international agreement.

On the other hand, as is well known, the European Union countries are important players as both carrier and cargo owner countries today, and their perspectives and decisions concerning the approval of the Rotterdam Rules is of very high importance for the future and international applicability of these rules.

In this context, this article will first focus on the history and the legal structure of the EU, and then study the European Union's stance on the Rotterdam Rules, the impact of the possible EU legislation preparation on the same areas, and the existent steps that are taken (as well as possible future steps) by the European Commission with regards to alternative legislation creation for the EU seas.

1 INTRODUCTION

Globally applicable rules have always been a necessity due to the merchant ships sailing through the territorial waters of different states and facing various problems generally stemming from the highly divergent judicial systems in these different geographies. In this context, to a great extent, the studies on this matter have concentrated on the liabilities of the carrier.

As is known, the current rules about the liabilities of the carrier in maritime transport came into force with the "Hague Rules" (1924 Brussels Convention) and Visby Rules (1968 Brussels Protocol).[1] The 1978 Hamburg Rules, prepared by United Nations, came into force in 1992 but the implementation couldn't reach the level of Hague Rules, which are implemented world-wide.

In this framework, the attempts for implementation of new international rules, namely Rotterdam Rules, and expansion of the liabilities of the carriers in maritime transport (though, not as much as Hamburg) and transferring some of the non-maritime transport liabilities to the carrier, have speeded up.

As a result of these attempts, on 11 December 2008, the United Nations General Assembly, in its 63rd session, adopted the "Convention of Contracts for the International Carrying of Goods Wholly or Partly by Sea". The signing ceremony was held in Rotterdam on 23 September 2009. Rotterdam Rules composed of 96 articles and 18 chapters, basically expanded the liabilities of the carrier compared to the Hague Rules, and expand carrier's liability to the entire carriage process, considering it as a part of the combined transport.

When analyzed thoroughly, by taking into account its role in the world trade and the global logistics services, the European Union is one of the mostly affected regions by these kinds of regulations with respect to its position as a block of states composed

---

[1] Actually, there is also a 1968 Special Drawing Right protocol amending the rules.

of both cargo owners and carriers. So, the analysis of the member states' views about the Rotterdam Rules and the attempts for probable alternative rules is important. Currently EU has set free its member states in adopting the Rotterdam Rules.[2] But the European Commission is preparing some binding regulations for member states and the European carriers are putting pressure on the Commission about this matter.

In this context, in the following sections, firstly EU's historical background and binding nature of the EU law for the member and candidate states will be briefly analyzed (regional law), then international regulations regarding the liabilities of the carrier will be summarized (international law) and finally the effects of EU's possible future legislation preparation about the liabilities of the carrier on the existing international law will be examined (the relationship between regional and international law).

2 THE HISTORICAL BACKGROUND OF THE EU AND THE EU LAW

Many political leaders have tried to create a united and powerful Europe in the history. But a real European integration movement, depending on the free-will of the individuals, could only start after the Second World War.[3] In this context, in the post-war period, the Western Europe states took the first step for the integration. France and England made an alliance with the Treaty of Dunkirk. In March 1948, these two powers and the Benelux countries, signed the Treaty of Brussels, later called as the Western European Union.

In those years, the USA, supporting the Europeans to act together as a single body, played an important role in the start of the integration process. In 1947, Harry Truman, the president of the USA, proposed an aid program, implemented under the leadership of George Marshall, the secretary of the state, to relieve Europeans from the hardships they were facing.[4] The European federalists aimed to establish a United States of Europe, and defended that the integration had to depend on the free will of the pub-

lic.[5] For this goal, at the end of a conference in London, they established The Council of Europe in Strasbourg, which would harmonize the laws of European states and promote the development of human rights and cultural cooperation in Europe.[6] The other positively effecting factor in the integration of Europe during those days was the establishment of North Atlantic Treaty Organization (NATO) in 1949. By means of this organization, the European powers left the critical defense issues to the NATO and concentrated on the economic development, cooperation and formation of the common law.[7]

However, the really successful steps, promoting further political integration in Europe, would be taken in the following years. In this context, the European decision makers concentrated on the energy sources (the most valuable one being coal during those days) and raw materials (iron and steel being at the top of the list).[8] European leaders, which came to conclusion that the political integration would only be realized through technical steps and prior economic integration, thought that the single market and the integrated European economy would be the catalyst for the solutions of political problems of the continent.[9]

This plan was announced by the Schumann Declaration on 9 May 1950, and the European Coal and Steel Community (ECSC), the first organization of the European integration was established in 1952. Germany, France, Italy, Belgium, Luxemburg and Netherlands were the first six signatories of the Treaty. The establishment of the European Economic Community (EEC) and European Atomic Energy Community (EAEC) by Treaties of Rome took the integration idea further and expanded it to other areas. Following the economic integration theories of the period, the free movement of the goods, services, capital and labor were seen as the main tools of establishment of single market in Europe and this was clearly expressed in the EEC Treaty.[10] The Single European Act (SEA), signed in 1986, finalized the steps of forming a single market by assigning a schedule, and finally, the single market has been established in 1993. Maastricht Treaty (Treaty of European Union (TEU), signed in 1992, had played a key role in transforming EEC into EU, and additionally has founded the three important pillars – European Communities (EC), Common Foreign and Se-

[2] The EU has many regional regulations for transport in general, and for maritime transportation in particular. These mentioned regulations have a binding character for the member states but do not have a regulatory character for the private law relationships between parties. The EU regulations concerning the carrier's liability details the carrier's liability epecially in air and road transport. The regulations related to the carrier's liability in maritime transportation are about the rights of the passengers and their luggage safety, yet, there are not detailed regulations pertaining maritime carriage of goods. See http://eur-lex.europa.eu/Result.do?idReq=1&page=1 ve http://eur-lex.europa.eu/Result.do?idReq=1&page=2

[3] For the historical analysis of the developments and integration efforts in Europe See: **Palmer, Robert Roswell (Edt.)**, *A History of the Modern World*, New York, 1995, p. 97-389.

[4] **Dinan, Desmond**: *Europe Recast: A History of European Union*, London, 2004, p.13-45.

[5] **Oudenaren, John Van**: *European Integration: A History of the Three Treaties'*, **Tiersky, Ronald (Edt.)**, Europe Today, Oxford, 1999, p.241-273.

[6] **Dinan**, p. 24-25.

[7] **Dinan**, p.15.

[8] **Dedman, Martin J.**: The Origins and the Development of the European Union 1945-1995: A History of European Integration, London, 1996, p.57.

[9] **Dinan**, s. 13-45.

[10] **Rosamond, Ben**: *Theories of European Integration*, New York, 2000, p. 50-73.

curity Policy (CFSP), Justice and Home Affairs (JHA) – on which the EU is built on. By Maastricht Treaty, an economic and monetary union (transition to Euro) policy has also been established. Amsterdam Treaty, signed in 1997, merged the existing legal texts and formed a legal framework for the union. Subsequently, Nice Treaty, came to force in 2003, replaced Amsterdam Treaty as the highest legal text of the EU. A probable constitution would play a key role for the EU integration to gain a legal identity. However, because of the lacking consensus on the matter (especially due to vetoes of the France and Netherlands), it was greatly simplified and has come to life with the Lisbon Treaty in 2007. This treaty has come to force in the end of the 2009 after several referendums and debates in member states.

During this historical process in Europe, the traces of the transformation from intergovernmental nation states relationship to the multi-level governance (local, national, supra-national levels jointly producing common policies) can be found. Today, the political power of the EU organs has greatly increased and this situation can be seen when highly developed legal framework of the EU – Acquis Communautaire – is examined.

In addition to the sui-generis "deepening axis", the European integration has also "widened" in time. The number of the EU members reached to 9 by the memberships of United Kingdom, Ireland and Denmark in 1973 and by the full membership of Greece, the total number of the integration movement reached to 10, starting the expansion to the southeastern Europe. The number of the EU members reached to 12 by the memberships of Spain and Portugal in 1986, and it reached to 15 by the memberships of the Finland, Sweden and Austria in 1995. After the end of the Cold War and the collapse of the Berlin Wall, the probability of the expansion of the Union towards east including Central and East European countries, and also unification of West and East Germany became a hot issue. After the unification of West and East Germany, by the fifth enlargement wave, Poland, Check Republic, Slovenia, Slovakia, Hungary, Estonia, Latvia, Latonia, Cyprus (de facto: South Cyprus) and Malta became the members of the EU in 2004 and the number of the members reached to 25. By the memberships of the Bulgaria and Romania in 2007, the number of the EU members has reached to 27. By the future membership of the current candidate countries; Turkey, Croatia, Macedonia, Iceland and Montenegro, the number of the EU members will reach to 32 and the remaining Balkan states will be the potential candidates of the future enlargement waves.

EU, during the above summarized deepening and widening processes, has developed a continuously evolving sui-generis supranational law.[11] Actually,

the EU Law (Acquis Communautaire), developed on the Law of Causality, different from the Case Law, is published within the in the Official Journal of European Communities, which today is composed of more than 100.000 pages including binding regulations for the member states. These EU regulations have started to affect maritime industry in time. For example, the single cabotage for EU and its related regulations has come to force during the last decades.

These new supranational regulation are indeed harmonious with the global regulations (for example International Maritime Organization (IMO) rules), but it takes them further for the EU member states and bring new standards (for example; European Maritime Safety Agency (EMSA) inspecting the maritime training in the member states, prevention of marine pollution by means of EU legislation supplementing the MARPOL Convention), and has brought new additional regional and binding rules for maritime transportation.

Likewise, probable EU legislation which will develop in the same areas covered by the Rotterdam Rules will cause a multiple law order in related fields.[12] Additionally, contrary to the voluntary Rotterdam Rules, the EU legislation will absolutely be binding for the member states unless they declare that they will not participate in and they can manage to stay out of the scope of the legislation by means of some political maneuvers, which the EU law permits.

So the innovations brought by Rotterdam Rules about the liabilities of the carrier will cause interceptions or contradictions between the developing and the current EU regulations, which are effective in the EU waters. Under these circumstances, the EU member states will implement both international and EU law on this matter.

As is known, in international trade, the existence of uniform rules and the successful implementation of these rules help to found an effective and global trade system and provide safe and secure trading activities. The carriage operations of goods, which are the subjects of the international trade, are very important for the international trade. These carriage operations are to a great extent done by means of maritime transportation (approximately 80 %).[13] Therefore, one can say that maritime transportation plays a key role in international trade.

In this context, in the following section, the current international regulations about the liabilities of the carrier (also affecting the EU member states) and the improvements which have been brought by Rotterdam Rules to these regulations will be examined. Subsequently, the EU's perspective on these regulations (the inter-connection of the regionally binding

---

[11] **Dinan**, p. 266-283.

[12] For the related EU regulation See: reference 2
[13] UNCTAD, 2008 Review of Maritime Transportation.

EU law and the globally binding international law) and the probable results of the development EU Acquis in the same fields will be examined.

3 THE CURRENT INTERNATIONAL REGULATIONS ABOUT THE CARRIER'S LIABILITY AND THE ROTTERDAM RULES

The international characteristic of the maritime sector in general and the maritime carriage of goods in particular, require uniform legal regulations in this field. This necessity, in the 20th century, has led to the preparation and the implementation of Hague, Hague-Visby and Hamburg rules (Conventions). These mentioned conventions aimed at reaching an international law order binding for the signatories in maritime carriage of goods.[14] However these conventions avoided of regulating all the issues related to maritime transport and solving all the problems of the sector, so the ownership, registry and possessor of the ship and the agent services issues were left to the national laws. Hague, Hague-Visby and Hamburg Conventions, basically, considered and regulated the loss and damage of the goods carried by sea, conditions of the irresponsibility from loss and damage and limitations of the liabilities.[15]

Also, the amount of the goods carried by sea increased due to the boom in international trade especially in the last decades. The increase in maritime transport traffic caused new legal problems.[16] Because of the development of new transport methods; construction of new type of ships for the new type of cargo, combination of maritime transport with other transport modes[17] and the innovations in the delivery methods of the international trade, the rules of these conventions had to be changed and adapted to new conditions. Also, the development of the e-trade and the replacement of the current written/printed papers of transport law by the papers prepared in electronic format[18] required the regulation of these new issues which were not regulated by the related conventions in the past.

The necessity of simplification of legal language and eliminating the vague expressions, the aim of taking different legal systems into a common position, the inclusion of the practical solutions of the implementation problems to the legal texts, constituted a basis for preparation of a new regulation concerning the international carriage of goods.

The preparation process of a new international convention started in 1990s by the joint study of The United Nations Commission on the International Trade Law (UNCITRAL) and the Committee Maritime International (CMI), and has been finally completed after a 10 year long extensive work. The Rotterdam Rules, which have appeared after this effort, have touched on the issues related to the transport law and in this framework (like the previous similar regulations) have not focused on the issues stemming from real law and agency law. The issues related to carriage are forming the basis of Rotterdam Rules but the issues like freight and right of claim are excluded. Rotterdam Rules have prepared the legal substructure of the e-trade as well as included regulations related to door-to-door and multimodal transport. The Rotterdam Rules have aimed at harmonizing the legislation about carriage of goods with containers in parallel to the development and increase in the container transport sector. The Rotterdam Rules have preferred to stay away from the doctrinal debates and showed a pragmatic approach. The current global implementation in maritime transport has been upgraded to an international convention for the first time by Rotterdam Rules and became binding for the signatory states. In order to eliminate the differences between the legal systems, Continental and the Anglo-Saxon laws have been harmonized in the text of the Rotterdam Rules.[19]

Article 1.5 of the Rotterdam Rules defines "carrier" as a person that enters into a contract of carriage with a shipper. To eliminate the confusions in practice, article 1.6 defines the concept of "performing party". Also, the definition of the performing party has been written in a quite wide perspective beyond the concept of carrier in practice. In the article 1.6 a person that performs or undertakes to perform any of the obligations with respect to receipt, loading, handling, stowage, carriage, care, unloading or delivery of the goods called performing party and will be responsible. This situation expands the definition and liabilities of the carrier written in the international regulations.

The general scope of application of the Rotterdam Rules is clarified by the article 5. According to this article, convention applies to contracts of carriage in

---

[14] Faria, J.A.Estrella: *"Uniform Law for the International Transport at UNCITRAL: New Times, New Players, New Rules"* (Texas International Law Journal, 2009, S. 44, p. 277-319).

[15] For detailed information, evaluation and comparisons See. Ilgın, Sezer: *"Hamburg Kurallarının Türk Taşıyan ve Taşıtanlara Etkisi (I)"* (Denizatı, 1993, S. 2-3, p. 45-48 ve Ilgın, Sezer: *"Hamburg Kurallarının Türk Taşıyan ve Taşıtanlara Etkisi (II)"* (Denizatı, 1993, S. 4-5, p. 37-44).

[16] Schelin, Johan: *"The UNCITRAL Convention on Carriage of Goods by Sea: Harmonization or De-Harmonization"* (Texas International Law Journal, 2009, S. 44, p. 321-327).

[17] Fujita, Tomotaka: *"The Comprehensive Coverage of the New Convention: Performing Parties and Multimodal Implications"* (Texas International Law Journal, 2009, S. 44, p. 349-373).

[18] Alba, Manuel: *"Electronic Commerce Provisions in the UNCITRAL Convention on Contracts fort he International Carriage of Goods Wholly or Partly by Sea"* (Texas International Law Journal, 2009, S. 44, p. 387-416).

[19] Durak, Onur Sabri: *Deniz Yolu ile Eşya Taşıma Hukuku'nda Son Gelişmeler ve Rotterdam Kuralları Üzerine Değerlendirmeler*, Yağız Muammer/Yılmaz, Ayşe (Edt.), TMMOB Gemi Makineleri İşletme Mühendisleri Odası IV. Ulusal Sempozyumu Bildiriler Kitabı (s. 94-106), İstanbul, 2009.

which the place of receipt and the place of the delivery are in different states, and the port of loading of a sea carriage and the port of discharge of the same sea carriage are in different states, but, according to the contract of carriage, any one of the following places should be located in a contracting state:
- The place of receipt,
- The port of loading,
- The place of delivery or,
- The port of discharge.

According to the article 5.2, the convention applies without taking in to consideration the nationality of the vessel, the carrier, the performing parties, the shipper, the consignee or any other interested parties. Article 6 of Rotterdam Rules designates the exeptions for the application of article 5. Article 6.1 states that the convention does not apply to the following contracts in liner transportation:
1 Charter parties; and
2 Other contracts for the use of a ship or of any space thereon.

As stated by the article 6.2, Rotterdam Rules do not apply to contracts of carriage in non-liner transportation except when:
1 There is no charter party or other contract between the parties for the use of a ship or of any space
   on it; and
2 A transport document or an electronic transport record is issued.

Obligations of the carrier are regulated in the 4th chapter of the Rotterdam Rules. According to the article 11, the carrier shall carry the goods to the place of destination and deliver them to the consignee in accordance with the terms of the contract of carriage. As stated by the article 12, the carrier or a performing party (the carrier in practice or any other person written in the contract) will have the responsibility of the goods in the period which starts at receiving the goods for carriage and ends when the goods are delivered. The parties may designate and extend the responsibility period with the contract in accordance with the Rotterdam Rules and limitations expressed in the convention.

By the article 13 of the Rotterdam Rules the specific obligations of the carrier has been designated. The carrier, during the period of its responsibility, shall properly and carefully receive, load, handle, stow, carry, keep, care for, unload and deliver the goods. Specific obligations of the carrier applicable to the voyage by sea are written in the article 14. The carrier is bound before, at the beginning of and during the voyage by sea to exercise due diligence to:
1 a.Make and keep the ship seaworthy;
2 b. Properly crew, equip and supply the ship and keep the ship so crewed, equipped and supplied throughout the voyage; and

3 c.Make and keep the holds and all other parts of the ship in which the goods are carried, and any containers supplied by the carrier in or upon which the goods are carried, fit and safe for their reception, carriage and preservation.

A carrier or a performing party, as stated in article 15, may decline to receive or to load, and may take such other measures as are reasonable, including unloading, destroying or rendering goods harmless, if the goods are, or reasonably appear likely to become an actual danger to persons, property or the environment. In the frame of the article 16 of Rotterdam Rules: after the loading of the goods, the carrier or a performing party may sacrifice goods at sea when the sacrifice is reasonably made for the common safety or for the purpose of preserving from peril human life or other property involved in the common adventure.

Liability of the carrier for loss, damage or delay has been written in the 5th chapter and between the articles 17 and 23 of the Rotterdam Rules. Article 17 explains the basis of liability, article 18 regulates the liability of the carrier for other persons and article 19 regulates the liability of maritime performing parties. Article 20 points out the joint and several liabilities, article 21 defines delay. In the article 22 the calculation of compensation has been written. Article 23, at the end of the 5th chapter, considers the notice in case of loss, damage or delay. In the following part, the liabilities of the carrier stemming from the loss, damage or delay are examined.

If the claiment proves that the loss, damage or delay took place during the period of the carrier's responsibility, the carrier is liable for loss of or damage to the goods as well as for delay in delivery. However, the carrier is relieved of all or part of its liability if he can prove his absence of fault that the loss, damage or the delay has not caused by him or any persons referred to in article 18. The carrier is also relieved of all or part of its liability, if he proves that the events or the circumstances referred to in article 17.3 caused or contributed to the loss, damage or delay. Being different from Hague Rules, in Rotterdam Rules the carrier must prove his absence of fault or negligence in order to be relieved of all or part of its liabilities. "The navigation fault" in article 4.2.(a) of Hague Rules has not been stated in Rotterdam Rules, in other words, it has not regarded as a cause that the carrier is relived of all or part of its liability. There is a condition of irresponsibility in Rotterdam Rules that Hague Rules does not include, is the "reasonable measures to avoid or attempt to avoid damage to the environment" in article 17.3 (n).

If the claiment proves that the loss, damage or delay was probably caused by or contributed to by,
1 The unseaworthiness of the ship;
2 The improper crewing, equipping and supliying of the ship; or

3 The fact that the holds or other parts of the ship in which the goods are carried, or any containers supplied by the carrier in or upon which the goods are carried, were not fit and safe for reception, carriage and preservation of the goods, the carrier is also liable, notwithstanding article 17.3, for all or part of the loss, damage or delay.

As required by the article 18, the carrier is not only liable for its own acts and omissions but also is liable for the breach of its obligations caused by the acts or omissions of:

1 Any performing party;
2 The master or crew of the ship;
3 Employees of the carrier or a performing party; or
4 Any other person that performs or undertakes to perform any of the carrier's obligations under the contract of carriage, to extent that the person acts, either directly or indirectly, at the carrier's request or under the carrier's supervision or control.

Unless the opposite is agreed in favor of the shipper in the contract or there are causes eliminating the limited liabilities of the carrier, the liability of the carrier is limited to 100 sterling per part or unit according to Hague Rules. However, the weight unit of the goods and limitation of the liabilities as to weight was not mentioned by Hague Rules. By Visby Protocol, that amended Hague Rules, the problem stemmed from the expression of "unit" tried to be solved and the limit of the liability was fixed with a certain amount of per part or per each kilogram of grossweight, whichever amount is the higher. In 1979, the protocol which was adopted in Visby in 1968 was amanded and a new system related to the limitation of the liabilities was formed. As required by the rules known as Special Drawing Right (SDR), the limit of liability per part or unit was 667.67 SDR or 2 SDR per each kilogram of grossweight, whichever amount is the higher. As required by the Hamburg Rules, signed in 1978 and came into force in 1992, the limit of liability is 835 SDR per part or 2.5 SDR per grosskilogram, whichever amount is the higher.[20]

In the 12th chapter of Rotterdam Rules, article 59 regulates "limits of liabilities", article 60 regulates "limits of liability for loss caused by delay" and article 61 regulates "loss of the benefit of limitation of liability". As required by the article 59, essantially regulating the limits of the carrier's liability, the carrier's liability for breaches of its obligations is limited to 875 SDR per package or other shipping unit, or 3 SDR per kilogram of the gross weight of the goods that are the subject of the claim or dispute, whichever amount is the higher, except when the value of the goods has been declared by the shipper and included in the contract particulars, or when a higher amount than the amount of limitation of lia-

bility set out in the article has been aggreed upon between the carrier and the shipper. When compared with the previous Hamburg Rules, it is seen that Rotterdam Rules have increased the limits of liabilities from 835 SDR to 875 SDR per part or unit, and from 2.5 to 3 SDR per weight unit. Neither the carrier nor any of the persons referred to in article 18 is entitled to the benefit of the limitation of liability, if the claimant proves that the loss resulting from the breach of the carrier's obligation under the convention was attributable to a personal act or omission of the person claiming a right to limit done with the intent to cause such loss or recklessly and with knowledge that such loss would probably result.[21]

Judicial or arbitral proceedings in respect of claims or disputes arising from the breach of an obligation of the carrier, must be applied in 2 years after the day on which the carrier has delivered the goods or, in cases in which no goods have been delivered or only part of the goods have been delivered, on the last day on which the goods should have been delivered. This period, provided in article 62 shall not be subject to suspension or interruption. However, as required by article 63, the person against which a claim is made may at any time during the running of the period extend that period by a declaration to the claimant. This period may be further extended by another declaration or declarations.

4 CONCLUSIONS: EUROPEAN UNION AND ROTTERDAM RULES

Adoption of Rotterdam Rules by a greater number of countries is crucial for the increasing implementation of the rules internationally. At this point, it can be said that especially the carriers are not approaching to these rules positively, because of their expanding liabilities by Rotterdam Rules compared to the Hague-Visby Rules. Since Rotterdam Rules both expand the carrier's liabilities in certain areas and lessen them in other areas when compared with Hamburg Rules, the comparison of Rotterdam Rules with Hamburg Rules will be the subject of another academic study. In this context, this paper has concentrated on the differences between Hague Rules, which have an implementation in a larger geography, and the Rotterdam Rules. When these differences (expanding liabilities of the carrier) are considered it can be understood why the carriers stay away from the Rotterdam Rules.

Similarly, the EU member states as leading carrier countries are approaching to Rotterdam Rules cautiously. Although the EU set free its members to about the rules, it is observed that EU Commission

---

[20] For the detailed study and the evaluation See: **Ilgın, Sezer:** *Deniz Hukuku – II,* İstanbul, 2008, p. 42-44

[21] The evaluation and the critics regarding the liability of the carrier See: **Durak,** p. 98-104.

is under pressure of the union's carriers in order to bring alternative regulations regarding the liabilities of the carriers.

At this point, it can be said that the European carriers are not considering the Rotterdam Rules as a positive development, but they are pressing for alternative legislation, to be exempted from these regulations at least for the maritime operations performed in the EU seas, in which the Acquis Communataire is valid, in case Rotterdam rules find a global implementation area.

Yet, the European maritime carriers are also uncertain about this issue because with regards to the liabilities, they are not also approaching positively to the regional regulations because they increase the number of the legislations to be complied with. The national and regional attitudes of the EU countries, which are both carriers having maritime fleets, and shippers making industrial products, will affect the future of Rotterdam Rules.

Today, one can say that there is an uncertainty in the EU, which is a regional political and economic integration movement developing sui-generis law for its own region, about Rotterdam Rules and which legislation (international or regional) will be effective within the EU seas.

There are contradictory opinions set forth about the European carriers' tendencies about the Rotterdam Rules. In fact, at this point, one can say that the carriers or the shippers lobbying activities on the new legislation, will determine the result.

Yet, we can still say that the EU member states will be the leading actors in the future of Rotterdam Rules' global implementation with their globally dominant maritime transport companies such as MAERSK, Hapag Lloyd, Hamburg-Süd, etc. At this point, the attitudes of the interested parties such as the carriers and the shippers (and their countries) on this issue will determine the future of this international legislation. The European carriers may put more pressure on the EU Commission to prepare alternative regional regulations, if they are forced to adopt Rotterdam Rules in future because of the commercial obligations.

Also, as mentioned before, if Rotterdam Rules are not globally adopted, the maritime transportation companies, which are generally critical of regional regulations, may choose a way in line with the Hague, Hague-Visby or Hamburg rules, and resist a probable EU legislation. In the final analysis, one can say that, as it is expanding its legislative framework day by day, the EU may establish a uniform legislation for the liabilities of the carriers in EU seas, probably not in the near future but certainly in the long term.

REFERENCES:

Alba, Manuel: "Electronic Commerce Provisions in the UNCITRAL Convention on Contracts for the International Carriage of Goods Wholly or Partly by Sea" (Texas International Law Journal, 2009, P. 44).

Dedman, Martin J.: The Origins and the Development of the European Union 1945-1995: A History of European Integration, London 1996.

Dinan, Desmond: Europe Recast: A History of European Union, London, 2004.

Durak, Onur Sabri: Deniz Yolu ile Eşya Taşıma Hukuku'nda Son Gelişmeler ve Rotterdam Kuralları Üzerine Değerlendirmeler, Yağız Muammer/Yılmaz, Ayşe (Edt.), TMMOB Gemi Makineleri İşletme Mühendisleri Odası IV. Ulusal Sempozyumu Bildiriler Kitabı, İstanbul, 2009.

Faria, J.A.Estrella: "Uniform Law for the International Transport at UNCITRAL: New Times, New Players, New Rules" (Texas International Law Journal, 2009, P. 44).

Fujita, Tomotaka: "The Comprehensive Coverage of the New Convention: Performing Parties and Multimodal Implications" (Texas International Law Journal, 2009, P. 44).

Ilgın, Sezer: Deniz Hukuku – II, İstanbul, 2008.

Ilgın, Sezer: "Hamburg Kurallarının Türk Taşıyan ve Taşıtanlara Etkisi (II)" (Denizatı, 1993, P. 4-5).

Ilgın, Sezer: "Hamburg Kurallarının Türk Taşıyan ve Taşıtanlara Etkisi (I)" (Denizatı, 1993, P. 2-3).

Oudenaren, John Van: European Integration: A History of the Three Treaties', Tiersky, Ronald (Edt.), Europe Today, Oxford, 1999.

Palmer, Robert Roswell (Edt.), A History of the Modern World, New York, 1995.

Rosamond, Ben: Theories of European Integration, New York, 2000.

Schelin, Johan: "The UNCITRAL Convention on Carriage of Goods by Sea: Harmonization or De-Harmonization" (Texas International Law Journal, 2009, P. 44).

http://eur-lex.europa.eu/Result.do?idReq=1&page=1 (Online, 27.05.2010)

http://eur-lex.europa.eu/Result.do?idReq=1&page=2 (Online, 27.05.2010)

16. Maritime Law of Salvage and Adequacy of Laws Protecting the Salvors' Interest

F. Lansakara
Master Mariner. LLM (Maritime Law) Consultant JMC NAUTICAL PTE LTD. Singapore

ABSTRACT: Under maritime law salvage is encouraged and given priority with respect to salvage awards in many aspects such as maritime liens, leniency on salvors' negligence, right to limit liability and to the extend of departure from "no cure no pay principle" in the case of unsuccessful salvage but have saved the environment. These laws in favor of salvage affecting the salvors are not straight forward and have been criticized in some cases. This paper discusses the relevant maritime law principles under each circumstances criticism they face and legal remedies available to safeguard the interest of the salvors and access the adequacy of maritime laws pro-tecting the salvors interest.

1 INTRODUCTION

Salvage covers section of maritime law under assistance at sea and in port. Modern principles of maritime salvage law established in the early part of 19th century but the maritime salvage practice existed long before that time. Modern day maritime salvage constitutes on three basic principles: when there is an imminent danger at sea concerning a marine peril; the salvor voluntarily render a service and; upon successful completion he will be awarded the salvage taking into consideration all the relevant factors including the value of the property and the degree of risk he has taken. Having its roots in the law of equity[1] maritime salvage bears very peculiar set of laws quite different from others. The well known principle "no cure no pay" is one of the yard stick taken in determining the salvage award but exceptions to this also have developed that is to considers further recovering some of expenses that have reasonably incurred in cases of contribution to marine environmental protection even without the success of salvaging the whole property. Salvors priority in maritime liens, leniency on negligence observed by the courts, allowed for limitation of liability by existing limitation liability convention and even departure from the no cure no pay principle were allowed for salvors by 1989 International Salvage Convention but, these are not without controversies. Although today salvage mainly depends upon 1989 International Salvage Convention its interpretation depends upon the national courts thus need the un-derstanding of the interpretation under varies jurisdictions specially the English and the American. English law and the Lloyds Open Form played a historical role in developing salvage law on the other hand American salvage law have also shown steady progress.

2 LEGAL INSTRUMENTS

Many of the countries are signatories to1989 International Salvage Convention. About 58 contracting states representing 47% of the world shipping tonnage including Poland, UK, USA, China, and Greece are some of the countries. On the other hand Lloyd Open Form has incorporated the provisions of the convention to be applied contractually; therefore, even a non contracting state will abide by salvage convention's terms and condition contractually provided the parties have signed the relevant Lloyd Open form. By keeping the tradition paramount Salvage Convention described the reward shall be fixed with a view to encourage salvage operation[2] in addi-

---

[1] Set of legal principles under common law tradition

[2] Article 13 (1) The reward shall be fixed with a view to encouraging salvage operations, taking into account the following criteria without regard to the order in which they are presented below
(a) the salved value of the vessel and other property;
(b) the skill and efforts of the salvors in preventing or minimizing damage to the environment;
(c) the measure of success obtained by the salvor;
(d) the nature and degree of the danger;
(e) the skill and efforts of the salvors in salving the vessel, other property and life;
(f) the time used and expenses and losses incurred by the salvors;

tion to these the Limitation Liability Convention has the provisions for salvors to limit their liability[3] and also 1993 International Convention on Maritime Liens and Mortgages in force ratified by few countries reestablishes that salvage claim having priority above all other claims. On the domestic laws the English Supreme Court Act[4] consider the salvage claims attached to a maritime lien and the English courts recognizes the priority available to a salvage claim above all other maritime lien claims.

Although the United States of America has not ratified the Limitation Liability Convention it has its own statute on limitation they, instead of tonnage limitation take into consideration the value of the salved ship and the freight to be earned and base on these values the salvor may limit his liabilities to the salved.

Although English law widely in use other common law countries and USA can also be considered for forum for handling salvage claims. Lloyd Open From has provisions for English law but it can be altered to suit the circumstances.

3 SALVAGE CLAIMS AND LIENS

English law embraces the salvage claim as a maritime lien as described under English Supreme Courts Act states that under any contract in relation to salvage services whether covered under salvage convention or not. Under the English law salvage awards are given priority on liens. They consider among other things damage done by a ship, seamen's wages, masters wages and disbursements. The courts will determine the distribution of the funds in order of their priorities[5] but there are no strict rules of rankings.

1 Admiralty Marshal's cost
2 Claimant's cost
3 Maritime Lienees
4 Mortgagees
5 Other in rem claimants
   – When there are several salvors the last in time take priority.
   – When there are different category of maritime liens the salvor take priority

– When a claimant has a damage lien subsequent to lien that preserved the ship (salvage) the damage lien will take priority over salvage[6].

English law considers the extinction of maritime liens under the following circumstances: Immunity; delay of law suit; upon providing financial security by the defendant; establishment of limitation fund; wavier; destruction of property; Judgment on liability; Judicial sale and; sister ship arrest.

1993 International Convention on Maritime Liens and Mortgages came into force in Sept 2004. Few countries have so far ratified this convention they are: Indonesia, Ecuador, Estonia, Nigeria, Monaco, Russia, St Vincent and Grenadians, Spain, Tunisia, Ukraine and Vanuatu.[7] This convention contained the provisions in relation to maritime liens similar to that have generally accepted by major maritime nations and it also has the provisions that each state under its own law may grant maritime liens on a vessel to secure claims other than those generally recognized.[8]

Maritime liens on salvage claims are less controversial and, without conflict with other laws. They are recognized equally by international conventions and national laws.

4 SALVORS RIGHT TO LIMIT LIABILITY

Prior to 1976 Limitation Liability Convention, salvors had no right to limit their liabilities in cases of negligence or misconduct which blame on salvor their liabilities were unlimited for example the Tojo Maru Case in 1972[9].

1976 Limitation of Liability Convention, states that ship owners and salvors as defined, may limit their liability in accordance with rules of this convention for claims in respect of loss or damaged to property occurring on board in relation to salvage operation[10] subject to certain exceptions such as

---

(g) the risk of liability and other risks run by the salvors or their equipment;
(h) the promptness of the services rendered;
(i) the availability and use of vessels or other equipment intended for salvage operations;
(j) the state of readiness and efficiency of the salvor's equipment and the value thereof.
[3] 1976 Limitation Liability convention Article 1
[4] Section 20 (2) of English Supreme Courts Act 1981 (SCA1981)
[5] SCA 1981 S21(6)

[6] The case "Veritas" 1901 the vessel was safely towed by the salvors but unfortunately her engine failed and a second salvor assisted her to prevent her from sinking. During the operation the vessel came into contact with landing stage belonging to the Dock Board. The Board used its statutory powers to remove the vessel and claim against the ship in this case priority was given to the Boards claim against the ship before considering the salvage awards.
[7] Article 4 (1) (c) Maritime Liens - Claim for reward for the salvage of the ship
Article 5 - Priority of maritime liens; Article 15- Conflict of laws ; Article 16 - Extinction of maritime liens because of time limit
[8] Article 6 –Other maritime liens Each State Party may, under its law, grant other maritime liens on a vessel to secure claims…subject to condition which include time bar and rank below Salvage lien.
[9] Tojo Maru case salvors were not allowed to limit liability under the old system existed before 1976 Limitation of liability Convention.
[10] Article 2 (1) (a) claims in respect of loss of life or personal injury or loss of or damage to property (including damage to harbor works, basins and waterways and aids to navigation), occurring on board or in direct connection with the operation of the ship or with salvage operations, and consequential loss resulting …

gross negligence if proved limitations will not be allowed. This new development was a direct consequence of 1972 Tojo Maru case. Limitation Liability Convention, ratified by 52 States about 50% of world shipping tonnage and its latest protocol with higher amount of limits has ratified by 37 states about 42 % of the world shipping tonnage. 1996 protocol to Limitation Liability Convention provides an enhanced compensation regime compare to the former. USA is not a party to these conventions but thy have their own statutory provisions.

Inclusion of salvors for limitation of liability is a recent development in favor of their rights.

5 NEGLIGENCE OF SALVORS

Salvor taking the risk to save the property and during the salvage operation accidents occur due to salvors negligence or misconduct the question is to what extend these negligence or misconduct affect the salvage award and whether the ship owner can make a counterclaim for damages. Decisions on these were difficult because of the extra ordinary nature of the job and involvement of the high risk.

Although both UK and USA are the signatories to the 1989 Salvage Convention the interpretation of provisions on salvors negligence by the courts have been different. Under the convention the salvor shall owes a duty to the owner of the vessel or other property in danger to carry out the salvage operations with due care also to exercise due care to prevent minimize damages to the environment and take assistance when reasonably requested[11]. The salvor's negligence may also deprive him of whole or part of the award [12] the convention is silence on how they are measured and, it has left the decisions to national courts.

The case "Alenquer" courts description to what extend the leniency can be granted has a notable

value. A brief outcome of the English law cases described below.
Case 1947 The Delphinula (Court of Appeal)
The salvor guilty of misconduct reduction in salved value due to his misconduct was taken into consideration and also a counter claim or independent action.
Case 1955 The Alenquer
No salvage award was made but the damage claim had to be paid in full the judge adhered to the general principle and described why leniency cannot be applied here " when the their behavior is criticized contrary to public interest, the result of the courts decision as such to discourage salvors of taking unnecessary risks"
Case 1972 Tojo Maru[13] (House of Lords)
It was held that when the salvage operation is successful but there is negligence of the salvor in the case of successful salvage the owners can counterclaim damages from the salvor and the measure of damage is the difference between undamaged value of the ship base on "no negligence" of the salvor and damaged value of the ship and, the salvage award to be calculated base on undamaged value of the ship.

Finally the salvage award and owners counterclaim will set off against each other the balance will be due owner or salvor. It was also held that when there is no success in salvage there can be no counter claims as well.

This case has the highest authority the House of Lords however its calculation of the award taking into account undamaged value of the ship although there were "no negligence" has created a friction between it and "no cure no pay" principle. There are also other concerns on application of this case law to cases with two or more salvors with only one at fault and how it can affect the one who is not at fault.

With regard to Limitation of liability of the salvors the Court of Appeal in the above case held that limitation of liability can be applied before setting off owners counter claim.

When considering the salvors' negligence or misconduct the American method is different from the English courts. They categorize them as distinguishable damages and independent damages. Distinguishable damages means they inherit in the situation for example the Tojo Maru case was a distinguishable damage and the independent damages means they were caused independently by the salvaor any counter claim for damages the independent damages may only consider. Limitation of liability under the American statute is the damaged value of the ship plus the freight in the course of being earned.

---

[11]Article 8 - Duties of the salvor and of the owner and master
1. The salvor shall owe a duty to the owner of the vessel or other property in danger:
(a) to carry out the salvage operations with due care;
(b) in performing the duty specified in subparagraph (a), to exercise due care to prevent or minimize damage to the environment;
(c) whenever circumstances reasonably require, to seek assistance from other salvors; and
(d) to accept the intervention of other salvors when reasonably requested to do so by the owner or master of the vessel or other property in danger; provided however that the amount of his reward shall not be prejudiced should it be found that such a request was unreasonable.
Article 8 has been incorporated in LOF 2000 cl J
The consequences of salvors misconduct is laid down in conventions article 18
[12] Article 18 - The effect of salvor's misconduct
A salvor may be deprived of the whole or part of the payment due under this Convention to the extent that the salvage operations have become necessary or more difficult because of fault or neglect on his part or if the salvor has been guilty of fraud or other dishonest conduct.

---

[13] 1972 Tojo Maru collision accident in Persian Gulf salvors agreed to tow her to Kobe during the salvage operation, the salvors negligence caused an explosion and heavy damages to the ship. They however were successful in towing her to final destination. The owners counterclaim damages from salvor due to negligence.

Under the Llloyd Open Form 2000 the salvor is required to have observed best endeavors[14], there has been no definition of best endeavors it is commonly used in industry and widely known therefore best endeavor means Standard of reasonableness is that of a prudent sailor acting properly in the interest of salved property.

The law with respect to negligence and misconduct of salvors their interpretations by English Courts are conflicting and the American Courts interpretations much preferred with respect to preservation of salvors rights.

6 SPECIAL COMPENSATION

Special compensation was introduced in 1989 Salvage Convention to compensate the salvors if their salvage operation has contributed to protection of the marine environment even though they could not earn full or any salvage award.

Under the 1989 Salvage Convention if a salvor has carried out salvage operation in respect of a vessel which by itself or its cargo threatened damage to the environment and has failed to earn reward under article 13 he shall be entitled to special compensation from the owner of the vessel equivalent to his expenses as defined in article 14. This appear to be a another step to encourage the salvor for saving the environment but the calculation of the salvage expenses without considering the profits or bonuses turn out to be an unpractical one. The principle issue in the Nagasaki Spirit case was concern with the definition of expenses in Article 14(3) and, in particular, that part of it which refers to "fair rate for equipment and personnel actually and reasonably used in the salvage operation…[15]". The question was whether is it permissible to include a market or profitable rate, or whether the salvor was entitled to solely to reimbursement of expenditure. House of Lords delivering the judgment held that fair rate under article 14(3) meant fair rate of expenditure and did not include any element of profits. This draws strong reaction from the salvors and after lengthy discussions the marine salvage community arrived with the solution. This was a set of clause giving the basis for calculation of special compensation including bonuses under the guide lines set up by International Salvage Union (ISU) and clarifying other relevant criteria known as Special Compensation and Indemnity Clause (SCOPIC). The solution provided by SCOPIC is, the parties to a salvage contract may agree to incorporate SCOPIC into any LOF contract by reference, therefore contracting out of Article 14 of the Convention. Such contracting out is allowed under article 6 of the Salvage Convention[16]. The ship owners P&I clubs have agreed through a code of conduct (a gentlemen agreement between P& I Clubs and ISU) to provide financial security required for SCOPIC compensation by a standard guarantee form known as ISU5.

Special compensation available to salvors under the convention have faced with problems in practical application, SCOPIC so introduced is a contractual obligation and not a statutory one. Salvage convention compensation limits to apply if SCOPIC is not agreed.

7 SUMMARY AND CONCLUSION

Conflict of law with respect to maritime liens on salvage less likely priority for salvor has maintained throughout, but uncertainty exists in the English law with respect to claims on salvors negligence American law appears to be more settled on this regard. Limitation of liability conventions applied to salvors but under American law different limits will consider since they are not party to international Limitation Liability Conventions. Special compensation applicable to salvor under the 1989 Salvage Convention not practical therefore contracts shall insert SCOPIC clause in order for the salvor to get reasonable rate including bonus.

The adequacy of the laws protecting the salvors interest today depends upon international conventions and national laws. The existing set of maritime salvage laws covering the interest of the salvors with respect to limitation of liability and priority in maritime liens appeared to be adequate, with regard to judgment on salvors misconduct and negligence the existing laws are not universal therefore inadequate. Similarly, special compensation provisions which are available under the salvage convention do not encourage the salvor and seems inadequate unless SCOPIC is inserted.

---

[14] LOF 2000 cl A

[15] 14(3) Salvor's expenses for the purpose of paragraphs 1 and 2 means the out-of-pocket expenses reasonably incurred by the salvor in the salvage operation and a fair rate for equipment and personnel actually and reasonably used in the salvage operation, taking into consideration the criteria set out in article 13, paragraph 1 (h), (i) and (j).
Article 13 the Criteria for fixing the rewards
(h) the promptness of the services rendered;
(i) the availability and use of vessels or other equipment intended for salvage operations;
(j) the state of readiness and efficiency of the salvor's equipment and the value thereof.

REFERENCE LIST

[1] 1989 International Salvage Convention
[2] 1976 Limitation of Liability Convention
[3] 1996 Protocol to Limitation of liability Convention
[4] Limitation of Liability Act USA 46 U.S.C SS181 -196

---

[16] Article 6 - Salvage contracts. 1. This Convention shall apply to any salvage operations save to the extent that a contract otherwise provides expressly or by implication

[5] 1993 International Convention on Maritime Liens and Mortgages.

[6] English Supreme Courts Act 1981.

[7] Lloyd Open From 2000

[8] International Salvage Union Form ISU5 and SCOPIC clause

17. The Hong Kong International Convention for Safe and Environmentally Sound Management of the Recycling of Ships Hong Kong 2009

J. R. De Larrucea & C. S. Mihailovici
Ciencia i Enginyeria Nautiques, Universitat Politécnica Catalunya, Spain

ABSTRACT: The new international convention on ship recycling was adopted by IMO in May 2009 in Hong Kong, China, in accordance with IMO in December 2009. The Hong Kong International Convention for the environmentally sound management and security ships[1] recycling is to ensure that vessels are to be recycled once they have reached the end of its useful life not involve an unreasonable risk to human health and safety or the environment.

This new agreement includes: the design, construction, operation and preparation of ships to facilitate safe and environmentally sound recycling, without compromising the safety and operational efficiency of ships, the operation of ship recycling facilities in terms of Safety and Environment and the establishment of an enforcement mechanism for ship recycling, incorporating certification and reporting requirements.

The text has been developed over three years of preparatory work before the Conference, with contributions from Member States of IMO and non-governmental organizations and in cooperation with the ILO and the parties of Convention Basilea[2].

1 INTRODUCTION

The new international convention on ship recycling was adopted by IMO in May 2009 in Hong Kong, China.

The Hong Kong International Convention for the safety and environmentally sound management of ship recycling, aims to ensure that vessels are to be recycled once they have etched the end of its useful life not involve an unreasonable risk to human health and safety or the environment.

The text has been developed over three years of preparatory work before the Conference, with contributions from Member States of IMO and non-governmental organizations and in cooperation with the ILO and the Basel Convention Parties.

The Conference has considered two important rules of uniform law: the 1989 Basel Convention on Tran boundary Movement of Hazardous Wastes and their Disposal. New York: UN, 1989, and the Convention of the ILO / ILO, Safety and health in ship breaking: Guidelines for Asian countries and Turkey. Geneva: ILO, October 2003.

This new agreement includes: the design, construction, operation and preparation of ships to facilitate safe and environmentally sound recycling, without compromising the safety and operational efficiency of ships, the operation of ship recycling facilities in terms of Safety and Environment and the establishment of an enforcement mechanism for ship recycling, incorporating certification and reporting requirements.

The entry into force of this Convention will take place twenty-four months that at least 15 countries have ratified it, the sum of the fleet of these is at least 40% of GT of the world merchant fleet and the annual volume of barge recycling of these countries during the ten years preceding the entry into force not less than 3% of GT's merchant fleet of these countries.

2 GENERAL OBLIGATIONS AND DEFINITIONS

Article 1 of the Convention defines the obligations of each Member State: the obligation to assume all the means available to prevent, reduce, minimize and, wherever possible, eliminate accidents and other risks that affect health human and the environ-

---

[1] Official name in English: Hong Kong International Convention for the Safe and Environmentally sound recycling of ships, 2009

[2] Control of Transboundary Movements of Hazardous Wastes and their Disposal. Countries party to the Convention, see: http://www.basel.int/ratif/convention.htm

ment caused by the ship recycling, and improve safety on the ship, the protection of human health and the environment over the life of the ship.

Members should endeavor to cooperate with the purpose of effective implementation of the Convention and the development of technologies and practices that contribute to security and sustainable ship recycling.

Article 2 of the Convention provides all the definitions necessary for proper interpretation.

The following defines some of the most relevant concepts:

- Boat means any vessel operating or has operated in the marine environment, including submersibles, floating craft and floating platforms, including ships stripped of equipment or being towed.
- Hazardous material means any material or substance that may create a hazard to human health or the environment.
- Recycling of ships: the activity of a full or partial disarmament of a boat in a ship recycling facility to recover materials and components for a reprocessing and reuse, product managers and other toxic materials, and includes associated operations such as storage and treatment of components and other materials in situ, but not further processed or disposed of in other facilities.
- Installation of ship recycling area defined as place, yard or facility used for the recycling of ships.
- Recycling Company, the owner of the ship recycling facility or other organization or person who assumes responsibility for the implementation of the ship recycling activity from the owner of the facility for the recycling of ships and who on assuming such responsibility agrees to bear all the obligations and liabilities imposed by the Convention.

3 SCOPE

According to Article 3, the Convention shall apply, unless willing to be otherwise in the Convention, to:

- Vessels entitled to fly the flag of a Member State or operating under its responsibility.
- Ship recycling facilities operating under the jurisdiction of a member state. Not apply, in line with other IMO Conventions:
- Warships, auxiliary vessels or vessels owned or operated by a member state and used for non-commercial service.
- Vessels under 500 GT or ships operating solely in its life in waters under the sovereignty or jurisdiction of the country which is entitled to fly the flag.

With respect to those ships flying the flags of states not belonging to this Convention, member states apply the requirements of the Convention to ensure that they are not given anyfavorable treatment.

4 CONTROLS AND INSPECTIONS

4.1 Controls relating to the recycling of ships

Member State shall establish controls over:

- Vessels flying its flag or operating under their authority, which must comply with the requirements of the Convention and implementing the measures.
- Ship recycling facilities under their jurisdiction. Must meet the requirements of the Convention and enforce the measures.

According to Article 5 Member States must ensure that vessels flying its flag or operating under authority and subject to inspection and certification are inspected and certified in accordance with regulations of the Annex.

4.2 Control and listing of hazardous materials

In accordance with this Convention, each State shall:

- Prohibit or restrict the installation or use of hazardous materials listed in Appendices 1 and 2 of the Convention on ships under their flag or jurisdiction, whether in ships, ports, shipyards or offshore platforms.
- New ships carry on board an inventory, verified by the Administration or any person or body authorized, Hazardous Materials. This inventory will be specific to each vessel and indicate the amount and location. Existing ships must comply with all possible with this list no later than 5 years after entry into force of the Convention, or before going to scrap if before this period.
- Vessels that are to be recycled only be recycled Ship Recycling Facilities authorized by the Convention and will be one in which doing a Ship Recycling Plan. Before entering the waste loading facilities, fuel and waste on board should be the minimum. Fuel tanks and cargo tanks that have contained any toxic or flammable substance shall be designed to enter and / or work in them.
- A Ship Recycling Plan must be carried out by the Facility where it will be recycled boat before starting any recycling process taking into account the guidelines developed by the Organization and the information provided by the owner of the boat. Information on the establishment, maintenance and monitoring of working conditions and the amount and type of hazardous substances to be treated, including those listed in the Inventory of Hazardous Materials.
- Vessels must pass an initial review before being put into service or before the International Certificate in Hazardous Materials is issued. In addi-

tion, inspections at intervals as the Administration but not exceeding 5 years. If repairs are carried out or any significant change in the structure will pass a special review to ensure that it continues to comply with the provisions of the Convention.

4.3 *Approval of ship recycling facilities*

Under Article 6 Member States must ensure that ship recycling facilities operating under its jurisdiction and that ships recycled to which this Convention applies to them, or boats treaties under Article 3.4 of this Agreement, are authorized according to the provisions of the Annex.

4.4 *Controls recycling facility boats*

- Each State should establish legislation, regulations and standards necessary to ensure that ship recycling facilities are designed, constructed and operated in a safe and environmentally sustainable under the regulations of this Convention.
- The Ship Recycling Facilities authorized by a State must establish procedures and techniques that do not endanger workers or contaminate the surroundings. Must prevent, reduce and / or minimize adverse environmental effects taking into account the guidelines developed by the Organization. Furthermore, only accept ships that comply with the provisions of this Convention and have all the documentation and certificates available.

4.5 *Charter plan recycling facility*

The ship recycling facilities must have a plan which includes:
- A policy ensuring the safety of workers and protection of human health and the environment,
- Identification of tasks and responsibilities of workers and managers in their tasks,
- Inform and prepare workers for safe and environmentally friendly facilities,
- Emergency response plan,
- Monitoring plan of the ship recycling,
- Report of discharges, emissions and accidents causing damage or potentially hazardous to workers and the environment
- Report of illness, accident, injury or other dislikes that may occur to workers.
It also establishes that each Ship Recycling Facility must have procedures to:
- Avoid explosions, fires, hazardous atmospheres and other risks for working with high temperature,
- Prevent spills and releases that could harm human health and the environment.
Regarding the handling of dangerous substances all Ship Recycling Facility must ensure that hazard-

ous materials listed in the inventory must be identified, labeled, packaged and handled by operators trained and equipped for it taking into account the guidelines of the Organization, in particular:
- Liquid hazardous waste and sediment
- Substances and objects with heavy metals like lead, mercury, cadmium and chromium,
- Highly flammable paints containing lead,
- Asbestos and materials containing asbestos,
- Plastic contaminants
- Products with CFCs
- Other hazardous materials that are not mentioned above and as part of the ship's structure.
Establishing a plan to respond to emergencies, which include:
- Ensure that equipment and procedures used and followed properly to protect the integrity of individuals and to avoid contamination of the environment,
- Providing a center for medical and first aid to the installation, as well as means of fire fighting, evacuation and prevention of pollution, and
- To inform and train workers in the Installation in accordance with its powers, including simulation exercises.
Facility workers should be trained to safely perform the tasks they are intended, and the proper use and maintenance of personal protective equipment. These teams will consist of:
- Protect your face and eyes,
- Protection of hands and feet
- Hearing protection, protection against radioactive contamination,
- Fall protection, and
- Appropriate clothing to their task.

4.6 *Information exchange*

According to Article 7 Member States which have authorized facilities for the recycling of ships provided to the Organization and other members, if required, relevant information, under the Convention, by which its authorization decision is based. This information will be shared with the shortest possible.

4.7 *Inspection of boats*

Article 8 defines where, by whom, when and what will be inspected the ship:
- Any vessel to which it is subject to this Agreement may be subject to inspection at any port or offshore terminal of another member state, by officers authorized by the member state. This inspection is limited to verifying that he is on board the International Certificate in Hazardous Materials Inventory.
- When a ship does not carry a valid certificate or are indications to believe that:

1 The condition of the ship or its equipment does not correspond to the particularities of the certificate and / or Part I of the Hazardous Materials Inventory, or

2 Board has not implemented a process for maintaining the Inventory of Hazardous Materials;
It may make a detailed inspection by the guidance of the Organization.

5 VIOLATIONS, DISCLOSURE, TECHNICAL ASSISTANCE AND COOPERATION

Articles 9 and 10 establish the modus operandi of the States in case of infringement. Member States should cooperate in the detection of violations, both on ships and in ship recycling facilities and the implementation of this convention.

Any violation of the requirements of this Convention within the jurisdiction of a Member State shall be prohibited and appropriate sanctions will be carried out according to the law itself. When a breach occurs the Member State must act according to its own laws or facilitate the administration of boat information and evidence in its possession which has violated the Convention.

In the case of a ship:
− The law shall be that of government where the offense.
− If the Administration is informed by another State of an offense, are investigating the incident and may request information and evidence to the State on the violation.
− The Administration shall inform the other State and the Organization of measures taken. These measures were taken during the period of one year and may be, warn, detain or refuse entry to its ports Boat.
− Avoid stopping or unduly delay a boat.
− When a ship is unduly detained or delayed, you will be compensated for any loss or damage.

In the case of a ship recycling facility.
− The law shall be that of the State having jurisdiction over the installation.
− If the State is informed by another State of an offense, it will investigate the incident and may request information and evidence to the other State.
− The State shall inform the other State and the Organization of measures taken.
These measures were taken during the period of one year.

According to Article 12 each State shall inform the Organization and the properly distribute the following information.
− List of ship recycling facilities and licensed jurisdiction.
− Contact details of the authorities, including the head.

− List of recognized organizations and inspectors working on behalf of Directors and its responsibilities.
− Annual list of vessels flying the flag of a State to which they have been issued the Certificate of Recycling, including the name of the recycling company and the location of the ship recycling facility set forth in the certificate.
− Annual list of the ships recycled within the jurisdiction of a State.
− Information on violations of the Convention.
− Actions on ships and recycling facilities under the jurisdiction of a State.

According to Article 13, States, directly or through the Organization or other international bodies will support those states requiring assistance regarding security and sustainable ship recycling:
− Personnel training
− Availability of technology, equipment and facilities,
− Research and joint development programs, and
− Effective implementation of the Convention and the guidelines developed by the organization.

6 CONCLUSIONS

The Basel Convention published the 'Technical Guidelines for the Environmentally Sound Management of the Full and Partial Dismantling of Ships' in 2003 which was a predecessor to the Hong Kong Convention. This document defined the 'environmentally sound management of ship dismantling' as: "taking all practicable steps to ensure that hazardous wastes or other wastes are managed in a manner that will protect human health".[3]

The Hong Kong Convention is in place to implement this through legislation and includes the presence of hazards on ships "hazardous substances such as asbestos, heavy metals, hydrocarbons, ozone-depleting substances and others"[4]

Secretary for Transport & Housing Eva Cheng said: "This significant international convention provides a single regulatory platform needed to address safety, health and environmental issues in the disposal of end-of-life ships. It will help protect the health of workers in recycling yards, reduce damage to the environment and be instrumental to the sustainable development of the shipping industry worldwide."

One of the most important requirements under the new convention is the need to have an updated inventory of hazardous materials aboard a ship so precautionary measures can be taken to protect workers and the environment.

---

[3] Technical Guidelines for the Environmentally Sound Management of the Full and Partial Dismantling of Ships
[4] http://www.imo.org

REFERENCES:

IMO International Convention for Safety Hong Kong and Recycling Boat environmentally reasonable. London: IMO, May 2009.

IMO London Convention 1972 and the 1996 Protocol on the Prevention of Marine Pollution by Dumping of Wastes and Other Matter (edition 2003). London: IMO, 2003.

IMO, MARPOL 73/74, Regulation 13G of Annex I. Revised edition 2003. London: IMO, December 2003.

UN, the 1975 Barcelona Convention for the protection of the marine and coastal Mediterranean region. New York: UN, 1975.

UN, 1989 Basel Convention on Tran boundary Movement of Hazardous Wastes and their Disposal. New York: UN, 1989.

ILO Safety and health in ship breaking: Guidelines for Asian countries and Turkey. Geneva: ILO, October 2003.

18. Maritime Delimitation in the Baltic Sea: What Has Already Been Accomplished?

E. Franckx
Research Professor, Member of the Permanent Court of Arbitration
President of the Department of International and European Law, Brussels, Belgium

ABSTRACT: To write a legal paper for an audience consisting primarily of experts in the field of navigation, transport, ocean engineering and maritime technology is not an easy task. Finding an appropriate box to tick when having to indicate the topic of the contribution when the title of the present contribution was submitted to the organizers of the conference, represented already a first hurdle. In the list of about 90 possibilities, not one really fitted the subject matter of the present contribution. This was particularly worrisome, because the instructions here read: "Choose maximum three topics". Consequently, I would particularly like to thank the organizers of the conference for having stretched somewhat the purview of the conference in order to accommodate a legal paper. The next difficulty, of course, rests on the shoulders of the present author, for he will have to write a legal paper understandable to an audience of which not all members may be familiar with the international law of the sea. Having served as a regional expert on maritime delimitation with respect to the Baltic Sea in a world-wide project initiated by the American Society of International Law, and still ongoing today,[1] a good number of publications by the present author on this topic have appeared in legal journals or specialized books. Despite the normal practice in legal papers of making extensive use of footnotes, the present paper will only make use of a minimum number of other references, in order to enhance its readability for non-lawyers. Instead, it will provide a listing in annex of the writings by the present author on the subject to which the interested reader, wanting to find out more concrete guidelines and information, may readily turn.

The present paper, first of all, is not concerned with maritime law -- a term to be found in the above-mentioned list -- but with the law of the sea. These two concepts, even though they might have been confused in the past, are today clearly distinguished from one another. The law of the sea concerns the rules of international law governing the different maritime zones and the activities carried out there. It forms a branch of international law, for it concerns mainly the relationship between states and is consequently also sometimes called "international law of the sea" or even "international public law of the sea". Maritime law, on the other hand, is part of the national law of a state, for it deals with private interests at sea in general, and with the relationship between those who exploit ships and those who make use of them more particularly. It is therefore sometimes also called "commercial maritime law".[2]

After having clarified a few crucial law of the sea notions, the present paper tries to bring some order in the maritime boundary agreements concluded so far in the Baltic Sea. It does so by taking a major political event as caesura, namely the disappearance of the former Soviet Union from the political map of the world during 1991. Finally, some concluding remarks will try to describe the present state of affairs, while comparing it to the situation as it existed on the eve of this major political event of 1991.

---

[1] The project started during the late 1980s and intended to provide an in-depth examination of the state practice arising from the more than 100 existing maritime boundary delimitation agreements already concluded at that time. The outcome of this project is reflected in the *International Maritime Boundary* series, published by Martinus Nijhoff, of which five volumes have already seen the light of day between 1993 and 2005. A sixth volume will appear in 2011.

[2] Salmon, J. (ed.) 2001, *Dictionnaire de droit international public*, Bruylant, Brussels, pp. 375 and 389. In English this latter branch of law is called "Admiralty Law".

1 INTRODUCTORY COMMENTS

In order to be able to tackle the issue of maritime delimitation according to the rules of the law of the sea, it seems appropriate to first introduce the reader to a number of maritime zones, as well as certain rules applicable therein, of particular interest here.

At present, the law of the sea is codified by means of the United Nations Convention on the Law of the Sea.[3] This document, also called the Constitution for the Oceans,[4] is at present adhered to by all states surrounding the Baltic Sea, including the European Union with respect to those areas for which it is competent. This document creates a number of maritime zones of which the most important are, for present purposes at least, the territorial sea, the continental shelf and the exclusive economic zone.

The territorial sea is a zone where the coastal state exercises sovereignty as on its land territory, with certain exceptions such as the right of innocent passage for foreign vessels (1982 Convention, Arts 2-32). The continental shelf and the exclusive economic zone are both zones of functional jurisdiction, meaning that the coastal state does not exercise sovereignty over these maritime zones, but only certain sovereign rights. With respect to the continental shelf, these sovereign rights are primarily related to the living and non living resources of the sea-bed and subsoil (id., Arts 76-85), and with respect to the exclusive economic zone these sovereign rights are moreover related to those same resources of the superjacent waters (id., Arts 55-75). All these zones are limited in distance: The territorial sea to a maximum of 12 nautical miles (nm) (id., Art. 3), the continental shelf and the exclusive economic zone to 200 nm (id., Arts 76(1) & 57 respectively). Only the continental shelf can in certain locations extend beyond this latter limit (id., Art. 76(4-8)), but because of the restricted size of the Baltic Sea, this eventuality does not apply there.

All these maritime zones are measured starting from the so-called baseline. This can be either the normal baseline, meaning the low-water line along the coast as marked on large-scale charts officially recognized by the coastal state (id., Art. 5), or a straight baseline in case the coastline is deeply indented and cut into or if there is a fringe of islands located in front of it (id., Art. 7). Under certain circumstances, moreover, bays can be closed off by means of a straight baseline measuring not more than 24 nm (id., Art. 10). It should nevertheless be noted that the influence of straight baselines on maritime delimitation in general, and in the Baltic Sea more particularly, remains in most cases negligible

for it will rather be the salient features on the coastline which will resort effect in this respect.

Historically, the territorial sea preceded the continental shelf, and the latter in turn preceded the exclusive economic zone. The origin of the territorial sea predates the creation of modern international law and was for the first time codified in the 1958 Convention on the Territorial Sea and the Contiguous Zone.[5] The origin of the continental shelf is usually related to the Truman Proclamation of 1945. This concept was later cast in legal wording by the International Law Commission of the United Nations and incorporated in the 1958 Convention on the Continental Shelf.[6] The concept of the exclusive economic zone, in turn, is a creation of the negotiations leading up to the 1982 Convention, where it was for the first time codified.

Delimitation of these maritime zones, finally, is necessary either between adjacent states, or where the distance between two opposite states is less than two times the maximum extent of a particular zone. Given the fact that the Baltic Sea is nowhere more than 400 nm wide, coastal states are obliged to conclude delimitation agreements not only with adjacent, but each time also with opposite states. The rules applicable to delimitation of these different zones are not identical, and have sometimes evolved over time between the United Nations codification efforts of 1958 and 1982. As far as the territorial sea is concerned, the rules remained identical and consist of three elements: 1) agreement; 2) if no agreement proves possible, the median line every point of which is equidistant from the baseline becomes applicable; 3) unless historic title or special circumstances demand a different delimitation line.[7] A similar rule existed in 1958 with respect to the delimitation of the continental shelf, the only substantial difference being that historic title disappeared as possible exception to the application of the equidis-

---

[3] United Nations Convention on the Law of the Sea. Multilateral convention, 10 December 1982, United Nations Treaty Series, vol. 1833, 397-581. This convention entered into force on 16 November 1994 (available at <www.un.org/Depts/los/convention_agreements/texts/unclos/unclos_e.pdf). At the time of writing there were 160 states party to the convention, as well as the European Community. Hereinafter cited as 1982 Convention.

[4] Expression used by one of the founding fathers of the 1982 Convention. See 'A Constitution for the Oceans', Remarks by Tommy T.B. Koh, of Singapore, President of the Third United Nations Conference on the Law of the Sea (available at <www.un.org/Depts/los/convention_agreements/texts/koh _english.pdf>).

---

[5] Convention on the Territorial Sea and the Contiguous Zone. Multilateral convention, 29 April 1958, UNTS, vol. 516, 205, 206-224. This convention entered into force on 10 September 1964 (available at <untreaty.un.org/ilc/texts/instruments/english/conventions/8_1_1958_territorial_sea.pdf>). Hereinafter 1958 TZ Convention. Of all Baltic Sea coastal states at present, only Estonia, Poland and Sweden are not a party to this convention. Latvia and Lithuania acceded in 1992. The German Democratic Republic acceded in 1973, but Germany never became a party.

[6] Convention on the Continental Shelf. Multilateral convention, 29 April 1958, United Nations Treaty Series, vol. 499, 311, 312-320. This convention entered into force on 10 June 1964 (available at <untreaty.un.org/ilc/texts/instruments/ english/conventions/8_1_1958_continental_shelf.pdf>). Hereinafter 1958 CS Convention. Of all Baltic Sea coastal states at present, only Estonia and Lithuania are not a party to this convention. Latvia acceded in 1992. The German Democratic Republic acceded in 1973, but Germany never became a party.

[7] 1958 TS Convention, Art. 12(1) and 1982 Convention, Art. 15.

tant line.[8] However, this rule was not retained in the 1982 Convention, mainly because, as clearly indicated by the International Court of Justice in 1969, the application of the equidistance principle to convex and concave coasts may well lead to highly inequitable results.[9] Instead, the delimitation rule with respect to the continental shelf has been stripped of any substantial guidance criteria and only retains the obligation for parties "to achieve an equitable solution".[10] The delimitation rule with respect to the exclusive economic zone is identical to that of the continental shelf.[11]

This absence of any concrete guidance at present with respect to the rules of delimitation concerning the continental shelf and the exclusive economic zone has led to a marked increase in delimitation cases being submitted before the International Court of Justice or arbitral tribunals. This seems not abnormal, because states very often have totally different perceptions of what constitutes an "equitable solution" in a particular delimitation dispute they are confronted with. Nevertheless, it is to be noted that so far not one single maritime boundary within the Baltic Sea has been arrived at by means of such third party dispute settlement procedures. Instead, parties have up till now always managed to solve these issues by diplomatic means.

Finally, it can be added that there appears to be a natural tendency for states to conclude maritime boundary delimitation agreements first in areas where few so-called special circumstances are present. As a result, the maritime boundaries still to be concluded more often than not are characterized by the presence of such complicating factors. Islands are one such factor, especially when they are located far from the mainland, are small in size, or prove to be uninhabited. No fixed rules exist in international law with respect to the weight to be attributed to islands in general. Especially if one or more of the just-mentioned situations apply to islands, the exact weight to be given to them remains highly uncertain. Either full effect can be attributed to them, or no effect, or in fact any gradation in between, depending on what the parties, or a court or tribunal as the case may be, consider to be equitable under the specific circumstances of the case. If the median line were to be applied between two opposite states, full effect would entail that the median line is calculated starting from the island, whereas no effect would imply that the parties involved agree to draw a line every point of which will be equidistant from their respective mainland. In the latter case it is as if the island simply did not exist, at least not from a maritime delimitation point of view. As mentioned above, all gradations in between these two extremes are possible as well.

2   MARITIME DELIMITATION IN THE BALTIC SEA UNTIL 1991

Four different periods can be distinguished when trying to categorize the maritime boundary agreements concluded so far in the Baltic Sea region. Three of them relate to the period up to 1991.

A first period concerns the years 1945-1972. This period started with a territorial sea agreement concluded between Poland and the former Soviet Union in 1958 and ended with the conclusion between the same countries of an additional agreement concerning the continental shelf in 1969. This first period is characterized by the fact that all these agreements were concluded between former Eastern bloc countries, with the exception of the involvement of neutral Finland. This should not come as a surprise, for it provided several political advantages for the Eastern bloc on the international level, as for instance the 12 nm territorial sea claim of the former Soviet Union, explicitly imbedded in the 1958 agreement just mentioned, and the fact that the former German Democratic Republic, at a time that this country was not allowed to participate in the law of the sea negotiations and the ensuing 1958 conventions, was nevertheless able to claim a continental shelf of its own, which was at that time contested by the former Federal Republic of Germany.

A second period concerns the years 1973-1985. This period starts with the conclusion of the so-called *Grundlagenvertrag* between both Germanies in 1972, which made it possible for them to enter into formal treaty arrangements, including those relating to their maritime boundary in the Baltic Sea. This resulted in the conclusion of such an agreement in 1974 delimiting the waters of Lübeck Bay. Since both Germanies were also admitted as members to the United Nations in 1973, it moreover opened the door for the conclusion of agreements between the two blocs, a typical feature of the maritime delimitation agreements concluded during this period. In 1978, for instance, the former German Democratic Republic settled its continental shelf delimitation with Sweden. No matter how important all these agreements may have been from a political point of view, their importance from a maritime delimitation point of view remained rather slim. The 1974 agreement between the two Germanies only concerned a maritime boundary line of about 8 nm and the 1978 agreement between the former German Democratic Republic and Sweden was rather easy to determine, because of the absence of any special features in the area to be delimited, and rather short as well, for the delimitation line finally agreed upon only had an overall length of 29 nm.

---

[8] 1958 CS Convention, Art. 6(1-2).
[9] North Sea Continental Shelf Cases (Denmark v. Germany; The Netherlands v. Germany), 1969 ICJ Reports 3.
[10] 1982 Convention, Art. 83(1).
[11] *Id.*, Art. 74(1)

A third period concerns the years 1985-1990, *i.e.* until the dissolution of the former Soviet Union. This has by far been the most productive period as well as the most interesting one from a delimitation point of view. Indeed, during this rather short period, spanning only half a decade, more maritime delimitation agreements were concluded than during the 40 preceding years. It is also the most interesting one, for most of them delimit areas where islands or other special circumstances are to be noted. Some of them merely consolidated previously concluded agreements between the parties. It should not be forgotten that the notion of the exclusive economic zone had in the meantime become well-established.[12]   Some treaties consequently only brought some order in previously concluded agreements by broadening their application to the exclusive economic zone. The former Soviet Union concluded two such agreements in 1985 with Finland and Poland. But most of them added new segments in areas burdened by islands or other special circumstances. The 1984 agreement between Denmark and Sweden is a good example, for the Danish islands of Bornholm, Christianso and Fredrikso as well as the uninhabited Swedish island of Utklippan, all located in the Baltic Sea proper, complicated the delimitation negotiations. Even more problematic was the delimitation between Sweden and the former Soviet Union, because both parties had totally opposite opinions on the effect to be given to the Swedish islands of Gotland and Gotska Sandön, both located at some distance from the coast: According to Sweden full effect should be given to these islands, while according to the former Soviet Union they should have no effect at all for according to this country the median line was to be calculated starting from the Soviet and Swedish mainland instead.   After some 20 years of negotiations both parties finally reached a compromise agreement in 1988 in which the disputed area was divided by attributing 75 % to Sweden and 25 % to the former Soviet Union, compensated by reciprocal fishing rights in each other's zone so created in reverse percentages, *i.e.* 75 % Soviet rights in the Swedish part of the formerly disputed zone and 25 % Swedish rights in the Soviet part. In 1988 Denmark and the former German Democratic Republic reached an agreement, whereby the Danish islands of Lolland, Falster, Møn and Bornholm and the German islands of Rügen and Greifswalder Oie all influenced the final line agreed upon. When Poland and Sweden reached an agreement the year after, they encountered exactly the same problem as the one encountered between the former Soviet Union and Sweden, namely the exact

weight to be attributed to the island of Gotland. It was solved in a similar manner in the 1989 agreement as well, namely by attributing 75 % of the disputed zone to Sweden and 25 % to Poland. Atypical was the agreement concluded during this period between the former German Democratic Republic and Poland in 1989, for it totally disregarded a previously concluded continental shelf boundary when adapting this old 1968 line to new circumstances. This period can be concluded by mentioning the adoption of a first tri-junction point agreement in the Baltic Sea, reached between Poland, the former Soviet Union and Sweden in 1989 as well.

At the eve of the dissolution of the former Soviet Union, therefore, the delimitation of maritime zones in the Baltic Sea had reached a very advanced stage, not easily encountered in other regions around the world, even though some of them, like the North Sea, did not know of the political divide characterizing the Baltic Sea region in those days. Disregarding for a moment the remaining tri-junction points, it could be stated that at that time the only remaining boundary to be settled concerned the area south and southeast of the islands of Bornholm between Denmark and Poland.

3   MARITIME DELIMITATION IN THE BALTIC SEA SINCE 1991

A fourth, and at present last period concerns the years following the disappearance of the former Soviet Union from the political map of the world in 1991. Suddenly, the delimitation picture in the Baltic Sea became much more complex. Three new entities emerged, namely Estonia, Latvia and Lithuania, while the former German Democratic Republic disappeared. This raised two sets of new problems in the Baltic Sea region. First, a number of new maritime boundaries needed to be agreed upon, where none had existed before. Starting from the South, the maritime boundaries between Russia and Lithuania, Lithuania and Latvia, Latvia and Estonia, and finally Estonia and Russia needed to be agreed upon, for these water expanses had never really required any delimitation under a unified Soviet state. Only the first three have so far been signed by the respective parties, namely in 1997, 1999 and 1996 respectively. All of these agreements, except the one between Latvia and Lithuania, have moreover entered into force by now. Second, the much more delicate issue of state succession came to the fore. In the case of the reunification of Germany this did not create too much difficulty. One maritime boundary simply became superfluous, namely the 1974 agreement relating to Lübeck Bay, mentioned above. Germany moreover recognized the maritime delimitation treaties concluded by the former German Democratic Republic, even though the

---

[12] By coincidence, it was also in 1985 that the International Court of Justice stated that the exclusive economic zone had become part of customary law. See Case concerning the Continental Shelf (Libyan Arab Jamahiriya/Malta), 1985 ICJ Reports 13.

one with Poland caused particular concern in certain quarters because of the disputed land frontier on which it was based. The situation with respect to Latvia, Lithuania and Estonia was totally different because here the question arose as to the exact legal status of previously concluded maritime boundary agreements by the former Soviet Union after their regained independence. Examples of the latter are the agreements concluded between Estonia and Finland of 1996 and between Estonia and Sweden of 1998. In principle these agreements did not explicitly rely on the agreements previously concluded by the former Soviet Union, a non-negotiable demand of Estonia, but the delimitation line itself did not change, a policy strictly adhered to by the other countries involved. Finally, during this period two more tri-point agreements were concluded: One between Estonia, Latvia and Sweden in 1997 and one between Estonia, Finland and Sweden in 2001.

Before concluding this last period, brief mention can be made of the only remaining maritime boundary agreement of the first category of this fourth period awaiting conclusion, namely the one between Estonia and Russia. A treaty was concluded on 18 May 2005, but because of an introductory declaration attached to it by the Estonian parliament during the internal ratification process, the Russian side withdrew its signature to this document.

4 CONCLUSIONS

Like on the eve of the dissolution of the former Soviet Union, one could argue that the situation as it existed at that time has almost been restored at present, despite the fundamental changes which occurred in the wake of this event, namely the disappearance of one coastal state, the reemergence of three others, and the lessening of the fundamental divide between East and West. Indeed, today all countries surrounding the Baltic Sea are members of the European Union, with the exception of Russia which now only retains the control over the Kaliningrad region and the eastern part of the Gulf of Finland. The maritime zones generated by these coastlines are not very enviable for the former is concave in nature while the latter constitutes a *cul de sac*. Even though a few more tri-points have been agreed upon, the situation is still that the area south and southeast of Bornholm remains to be divided, be it that one has to add now that one maritime boundary is still awaiting (a second) agreement, namely between Estonia and Russia, and another one entry into force, namely between Latvia and Lithuania.

BIBLIOGRAPHY

Franckx, E. 1988, "'New' Soviet Delimitation Agreements with its Neighbors in the Baltic Sea', *Ocean Development and International Law Journal*, vol. 19, no. 2, pp. 143-158.

Franckx, E. 1989, 'The 1989 Maritime Boundary Delimitation Agreement Between the GDR and Poland', *International Journal of Estuarine and Coastal Law*, vol. 4, no. 4, pp. 237-251.

Franckx, E. 1990, 'Maritime Boundaries and Regional Co-operation', *International Journal of Estuarine and Coastal Law*, vol. 5, no. 1-3, pp. 215-227.

Franckx, E. 1990, 'First Trijunction Point Agreed upon in the Baltic Between Poland, Sweden and the U.S.S.R.', *International Journal of Estuarine and Coastal Law*, vol. 5, no. 4, pp. 394-397.

Franckx, E. 1991, 'International Cooperation in Respect of the Baltic Sea', in *The Changing Political Structure of Europe: Aspects of International Law*, eds. R. Lefeber, M. Fitzmaurice & E. W. Vierdag, Martinus Nijhoff, Dordrecht, pp. 245-277 (especially pp. 254-269).

Franckx, E. 1992, 'Maritime Boundaries and Regional Cooperation in the Baltic', *International Journal of Legal Information*, vol. 20, no. 1-3, pp. 18-23.

Franckx, E. 1992, 'EC Maritime Zones: The Delimitation Aspect', *Ocean Development and International Law Journal*, vol. 23, no. 2-3, pp. 239-258.

Franckx, E. 1993, 'Region X: Baltic Sea Maritime Boundaries', in 1 *International Maritime Boundaries*, eds. J. I. Charney & L. M. Alexander, Martinus Nijhoff, Dordrecht, pp. 345-368.

Franckx, E. 1993, [16 individual boundary reports (Report Numbers 10-1 until 10-12)], in 2 *International Maritime Boundaries*, eds. J. I. Charney & L. M. Alexander, Martinus Nijhoff, Dordrecht, pp. 1915-2104.

Franckx, E. 1995, 'The New United Nations Law of the Sea Convention and the Problem of the Maritime Delimitation in the Baltic Sea', in *Baltic Sea Yesterday, Today, Tomorrow*, ed. J. Bergholcs, Via Baltica, Riga, pp. 18-19.

Franckx, E. 1996, 'Finland and Sweden Complete Their Maritime Boundary in the Baltic Sea', *Ocean Development and International Law Journal*, vol. 27, no. 3, pp. 291-314.

Franckx, E. 1996, 'Frontières maritimes dans la mer Baltique : passé, présent et futur', *Espaces et Ressources Maritimes 1995*, vol. 9, pp. 92-115.

Franckx, E. 1996, 'Baltic Sea: Finland-Sweden Delimitation Agreement', *International Journal of Marine and Coastal Law*, vol. 11, no. 3, pp. 394-400.

Franckx, E. 1996, 'Maritime Boundary Delimitation in the Baltic', in *The Baltic Sea: New Developments in National Policies and International Cooperation*, eds. R. Platzöder & P. Verlaan, Martinus Nijhoff, The Hague, pp. 167-177.

Franckx, E. 1996, 'Maritime Boundaries in the Baltic Sea: Past, Present and Future', *Maritime Briefing*, vol. 2.

Franckx, E. 1997, 'Two New Maritime Boundary Delimitation Agreements in the Eastern Baltic Sea', *International Journal of Marine and Coastal Law*, vol. 12, no. 3, pp. 365-376.

Franckx, E. 1997, 'Les délimitations maritimes en mer Baltique', *Revue de l'Indemer*, vol. 5, pp. 37-76.

Franckx, E. 1998, 'Finland-Sweden (Bogskär Area) (Report Number 10-13)', in 3 *International Maritime Boundaries*, eds. J. I. Charney & L. M. Alexander, Martinus Nijhoff, The Hague, pp. 2539-2555.

Franckx, E. 1998, 'Baltic Sea Update (Report Number 10-14)', in 3 *International Maritime Boundaries*, eds. J. I. Charney & L. M. Alexander, Martinus Nijhoff, The Hague, pp. 2557-2573.

Franckx, E. 1998, 'Two More Maritime Boundary Agreements Concluded in the Eastern Baltic Sea in 1997', *International*

Journal of Marine and Coastal Law, vol. 13, no. 2, pp. 274-283.

Franckx, E. 1998, 'Maritime Boundaries in the Baltic', in *Boundaries and Energy: Problems and Prospects*, eds. G. H. Blake, M. Pratt, C. Schofield & J. A. Brown, Kluwer Law International, London, pp. 275-295.

Franckx, E. & Pauwels, A. 1998, 'Lithuanian-Russian Boundary Agreement of October 1997: To Be or Not To Be?', in *Liber amicorum Günther Jaenicke: zum 85. Geburtstag*, eds. V. Götz, P. Selmer & R. Wolfrum, Springer, Berlin, pp. 63-95.

Franckx, E. 2000, 'The 1998 Estonia-Sweden Maritime Boundary Agreement: Lessons to be Learned in the Area of Continuity and/or Succession of States', *Ocean Development and International Law Journal*, vol. 31, no. 3, pp. 269-284.

Franckx, E. 2000, 'Maritime Boundaries in the Baltic Sea: Post-1991 Developments', *Georgia Journal of International and Comparative Law*, vol. 28, no. 2, pp. 249-266.

Franckx, E. 2000, 'Maritime Boundaries in the Baltic Sea: Post-1991 Developments', *VLIZ Collected Reprints: Marine and Coastal Research in Flanders*, vol. 30, pp. 310-325.

Franckx, E. 2001, 'New Maritime Boundaries Concluded in the Eastern Baltic Sea Since 1998', *International Journal of Marine and Coastal Law*, vol. 16, no. 4, pp. 645-659.

Franckx, E. 2002, 'New Maritime Boundaries Concluded in the Eastern Baltic Sea Since 1998 (bis)', *International Journal of Marine and Coastal Law*, vol. 17, no. 2, pp. 263-266.

Franckx, E. 2002, [8 individual boundary reports (Report Numbers 10-15 until 10-21)], in 4 *International Maritime Boundaries*, eds. J. I. Charney & R. W. Smith, Martinus Nijhoff, The Hague, pp. 2995-3017.

Franckx, E. 2001, 'New Maritime Boundaries Concluded in the Eastern Baltic Sea Since 1998', *VLIZ Collected Reprints: Marine and Coastal Research in Flanders*, vol. 31, pp. 344-358.

Franckx, E. 2005, 'Region X: Baltic Sea Boundaries', in 5 *International Maritime Boundaries*, eds. D. A. Colson & R. W. Smith, Martinus Nijhoff, Leiden, pp. 3507-3535.

Franckx, E. & Kamga, M. 2008, 'L'existence éphémère du Traité de délimitation maritime entre la République d'Estonie et la Fédération de Russie en mer Baltique', *Annuaire du Droit de la Mer 2007*, vol. 12, pp. 393-423.

Franckx, E. 2010, 'Estonia-Russian Federation (Report Number 10-22)', in 6 *International Maritime Boundaries*, eds. D. A. Colson & R. W. Smith, Martinus Nijhoff, The Hague [forthcoming].

Ships Monitoring System; A Decision Support Tool

19. Ships Monitoring System

L. V. Popa

Navigation and Maritime Transport, Constanta Maritime University, Romania

ABSTRACT: Due to the increasing of the piracy attacks all over the world, the maritime transport is now facing increased risk and security problems. To prevent and minimize the impact of the piracy, the owners and the maritime administrations must take preventive actions/measures. One of these actions is the implementation at a global level of the Long Range Identification and Tracking system (LRIT). The system, mandatory under SOLAS Chapter V, Regulation 19-1, is now operational since one year and proved to be a useful security tool against piracy.

1 INTRODUCTION

Maritime piracy has been on the rise for years, according to the International Maritime Bureau's (IMB) Piracy Reporting Center. But until 2008, when pirates operating off the coast of Somalia hijacked a ship full of Russian war-tanks and an oil supertanker, the crime drew limited international attention. By early 2009, more than a dozen countries had deployed their navies to the Gulf of Aden to counter piracy, and the United Nations passed four resolutions in 2008 on the issue. In April 2009, stakes grew higher after the U.S. Navy killed three Somali pirates, and took one captive in the rescue operation of a U.S. cargo ship captain taken hostage. There are a range of measures available to combat piracy--from onboard defense systems to naval deployments to preemptive strikes. Pirate attacks are largely confined to four major areas:

– The Gulf of Aden, near Somalia and the southern entrance to the Red Sea;
– The Gulf of Guinea, near Nigeria and the Niger River delta;
– The Malacca Strait between Indonesia and Malaysia;
– The Indian subcontinent, particularly between India and Sri Lanka.

In 2008, maritime piracy reached its highest level since the International Maritime Bureau's Piracy Reporting Center began tracking piracy incidents in 1992. Global piracy increased 11 percent, with piracy in East Africa up a stunning 200 percent. Of the forty-nine successful hijackings, forty-two occurred off the coast of Somalia, including the capture of an oil supertanker, the Sirius Star. Five hijackings were off the Nigerian coast, though the IMB suggests attacks in that area are underreported. In other areas of the world, including Indonesia, piracy dropped.

The shipping industry has urged greater action on the part of the world's navies. But many ships are not even using basic deterrents.

There is no quantitative research available on the total cost of global piracy. Estimates vary widely because of disagreement over whether insurance premiums, freight rates, and the cost of re-routings should be included with, for instance, the cost of ransoms. Some analysts suggest the cost is close to $1 billion a year, while others claim losses could range as high as $16 billion. Some experts such as Martin N. Murphy, author of a 2007 study on piracy and terrorism, warn against exaggerating the threat posed by maritime pirates. He notes that even $16 billion in losses is a small sum in comparison to annual global maritime commerce, which is in the trillions of dollars.

2 MECHANISMS FOR COMBATING PIRACY

A range of options exists for combating maritime piracy, but experts stress that most of the current tactics are defensive in nature, and do not address the state instability that allows piracy to flourish. The mechanisms used or under consideration in the most prevalent piracy area, the Gulf of Aden, can be classified as follows:

2.1 Onboard deterrents

Individual ships have adopted different onboard deterrents. Some use rudimentary measures such as fire hoses, deck patrols, or even carpet tacks to repel pirates. Others use a nonlethal electric screen with a loudspeaker system that emits a pitch so painful it keeps pirates away. Most do not arm their crews, both because ship workers tend to be unskilled and because many do not want to carry weapons, fearing that pirates will target them if they are armed. The shipping industry has urged greater action on the part of the world's navies. But many ships are not even using basic deterrents, writes retired U.S. Navy Commander John Patch in *Proceedings* magazine. [6]

2.2 Naval deployments

By January 2009, an estimated thirty ships were patrolling an area of about 2.5 million square miles. More than a dozen countries--including Russia, France, the United Kingdom, India, China, and the United States--had sent warships to the Gulf of Aden to deter pirates. There were also two multinational anti-piracy patrols in the area: the European Union's military operation, called EU NAVFOR, which began in December 2008; and a multinational contingent, known as Combined Task Force 150, which was originally tasked with counterterrorism efforts off the Horn of Africa. The United States announced a new task force, CTF-151, in January 2009. Some analysts, including a blogger for the U.S. Naval Institute, suggest that the new task force will allow the United States to seek a non-Western approach to counter piracy by partnering with Eastern navies. [6]

2.3 Long Range Identification and Tracking

Experts unanimously stress that the only effective long-term piracy deterrent is a stable state. When Somalia was briefly under the control of the Islamic Courts Union in 2006, the piracy acts stopped completely. Until recently, sovereignty has prevented outside states from targeting inland pirate infrastructure. A UN resolution passed on December 2, 2008, allows states to enter Somalia's territorial waters in pursuit of pirates, and another resolution passed on December 16, 2008, implicitly authorizes land pursuit.

On 19 May 2006, the International Maritime Organization (IMO) adopted Resolutions of the Marine Safety Committee MSC 202 (81) and MSC 211 (81) which states amendments to the International Convention of Safety of Life At Sea, 1974 (SOLAS) and introduces the timely establishment of the Long-Range Identification and Tracking system (LRIT). [4]

A robust international scheme for long-range identification and tracking of ships is an important and integral element of maritime security. An active and accurate long-range identification and tracking system also has potential safety benefits, most notably for maritime search and rescue. Accurate information on the location of the ship in distress as well as ships in the vicinity that could lend assistance will save valuable response time to affect a timely rescue.

At the 83rd Maritime Safety Committee the purpose and scope of LRIT was extended ultimately to include safety and environmental protection applications.

The requirements concerning LRIT have been introduced into SOLAS, Chapter V ("Safety of Navigation"), Regulation 19-1. In accordance with Paragraph 8.1 of Regulation 19-1, "Contracting Governments shall be able to receive long-range identification and tracking information about ships for security and other purposes as agreed by the Organization". Such "other purposes" would for instance include Search and Rescue (SAR), as explicitly mentioned in the new SOLAS provisions, as well as maritime safety in general and marine environment protection purposes as agreed by Resolution MSC 242(83) adopted on 12 October 2007. The IMO LRIT requires that all passenger ships including high speed craft, cargo ships of 300 gross tonnage and above, mobile offshore drilling units should automatically transmit every 6 hours the identity of the ship, the position report and time of the position. [4]

Furthermore, IMO also adopted on 19 May 2006, Resolution MSC 210 (81) amended and modified by MSC 254 (83) which establishes performance standards and functional requirements for the LRIT of ships. This states that all LRIT Data Centers and the International LRIT Data Exchange should conform to functional requirements not inferior to those specified in the Annex to the Resolution. [4]

The performance standards were then revised through Resolution MSC 263(84) adopted on May 2008 - Revised performance Standards and functional requirements for the LRIT of ships (this revokes MSC 210(81), MSC 254(83)). The system specifies that 4 position messages per day are stored and available for those actors entitled to access the LRIT information. The international LRIT system receives, stores and disseminates LRIT information on behalf of all Contracting SOLAS Governments.

The LRIT system consists of the ship borne LRIT information transmitting equipment, the Communication Service Provider(s), the Application Service Provider(s), the LRIT Data Centre(s), including any related Vessel Monitoring System(s), the LRIT Data Distribution Plan and the International LRIT Data Exchange. [1]

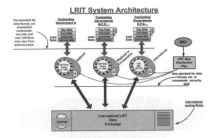

Fig.1. LRIT System Architecture

Certain aspects of the performance of the LRIT system are reviewed or audited by the International Mobile Satellite Organization (IMSO) appointed as LRIT Coordinator in December 2008 [MSC 275(85)].

Each Administration should provide to the LRIT Data Centre it has selected, a list of the ships entitled to fly its flag, which are required to transmit LRIT information, together with other salient details and should update, without undue delay, such lists as and when changes occur.

The obligations of ships to transmit LRIT information and the rights and obligations of Contracting Governments and of Search and rescue services to receive LRIT information are established in regulation V/19-1 of the 1974 SOLAS Convention.

It should be noted that regulation V/19-1.1 provides that:

Nothing in this regulation or the provisions performance standards and functional requirements adopted by the Organization in relation to the long-range identification and tracking of ships shall prejudice the rights, jurisdiction or obligations of States under international law, in particular, the legal regimes of the high seas, the exclusive economic zone, the contiguous zone, the territorial seas or the straits used for international navigation and archipelagic sea lanes.

3 SYSTEM ARCHITECTURE

3.1 *LRIT components*

The international LRIT system consists of:
- ship borne LRIT information transmitting equipment;
- Application System Provider(s) – ASP;
- Communication Service Provider(s) – CSP;
- National, Regional, Co-operative and International Data Centre(s) including related Ship Monitoring System(s) – SMS(s) and Vessel Traffic Service(s) – VTS(s); [1]
- International Data Exchange – IDE;
- the Data Distribution Plan - DDP; and
- LRIT Co-ordinator.

3.2 *Ship borne equipment*

Ship borne LRIT equipment should be capable to automatically transmit LRIT data to the selected LRIT Data Centre at 6-hour intervals and to be configured remotely to transmit data at variable intervals ranging from 15 minutes to 6 hours, following receipt of polling commands.

3.3 *Application and Communication Service Providers*

Application Service Providers (ASPs) provide services to the selected LRIT Data Centres and should:
- be recognized by the contracting governments of the associated Data Centre;
- provide a communication protocol interface between Communication Service Providers (CSPs) and Data Centres to enable remote integration of ship equipment into selected LRIT Data Centre and automatic management, configuration, modification, suspension and recovery of LRIT data transmissions;
- add defined set of data to each transmission of the LRIT information;
- provide an integrated transaction management system for the monitoring of LRIT data throughput and routine; and
- ensure that LRIT data is collected, stored and routed in a reliable and secured manner.

Communication Service Providers (CSPs) connect the ship-borne equipment with the ASP in order to ensure the end-to-end reliable, timely and secure transfer of LRIT data. Communication between ships and Data Centers may be secured by different Satellite and Terrestrial CSPs.

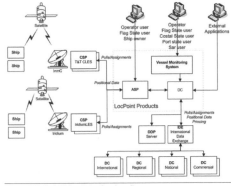

Fig.3. Global Ships Monitoring System

3.4 *International Data Exchange (IDE)*

The International LRIT Data Exchange is a message handling service that connects all LRIT Data Centres and route LRIT data between particular Data Centres using a standard agreed protocol, secure access and

routing table to establish the correct distribution of the reports. Additionally it should:
- use a store and foreword-buffer to ensure LRIT data is received;
- automatically maintain journal containing headers of all routed messages;
- archive journal for at least one year for invoicing and audit purposes; and
- not store or archive LRIT data.

3.5 Data Distribution Plan (DDP)

The DDP is the set of rules governing the distribution of the LRIT reports between the users of the system. The rules are established by each Contracting Government and uploaded accordingly on the DDP server hosted and maintained by the IMO.

3.6 LRIT Data Centre (DC)

Each SOLAS Contracting Government (CG) is required to establish or participate in a National/Regional/Cooperative Data Centre. Once the DC was established, all SOLAS ships under the flag of the relevant CG will report to the nominated DC.
The appointed DC / ASP will undertake in general the following tasks: integration of ship equipment into the designated DC, initial terminal compliance testing and certification in conjunction with the Ship Operator (or nominated regulatory representatives), management of the DC activities, connection of the DC to the wider international LRIT network via the International Data Exchange (IDE), and coordination of Data Centre-to-Data Centre billing arrangements.

4 THE EU LRIT DATA CENTRE

In line with IMO requirements, the European Member States have decided to establish an European Union Cooperative Data Centre (EU LRIT CDC).The objective of the EU LRIT CDC is the identification and tracking of EU Flagged ships and the integration in the wider International LRIT system. The main advantage is that all Member States can share a LRIT information repository, a common interface to the International Data Exchange (IDE) for requesting LRIT information on ships flying non-EU flags, and a common interface to LRIT information eventually via the Safe Sea Net system.

According to paragraph 1 of the Council Resolution, the Commission is in charge of managing the EU LRIT CDC, in cooperation with Member States, through the European Maritime Safety Agency (EMSA). The Agency is more particularly in charge of the technical development, operation and maintenance of the EU LRIT CDC. It also "stresses that the objective of the EU LRIT CDC should include

maritime security, Search and Rescue (SAR), maritime safety and protection of the marine environment, taking into consideration respective developments within the IMO context."

The EU LRIT CDC is operational since June 2009 in accordance with all IMO performance standards and requirements.

The general architecture of the EU LRIT system and the links between the EU LRIT Data Centre and other components of the system such as the links with the IDE, DDP, and EU LRIT Ship database are shown in the figure below. The components are similar to the International LRIT system and the EU Data Centre links with the IDE to obtain information from non-EU flagged ships.

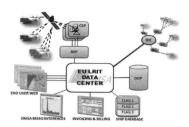

Fig.2. EU LRIT Data Centre

4.1 EU LRIT CDC and piracy

In order to assist the EU NAVFOR efforts in fighting the piracy acts off Somalia coastal area, the EU LRIT CDC has developed a specific anti-piracy tool based on the Flag State LRIT reports. The tool consists of a defined polygon off Somalia coastline where all EU ships entering the polygon automatically send an alert to EU NAVFOR and changed their reporting rate from the default 6 hrs to 1 hr (see Fig. 3). The EU NAVFOR has direct access to this tool and they can visualise and closely track each EU ships navigating in the area. The close monitoring rate provides a better coordination for the EU NAVFOR escorting ships in the area.

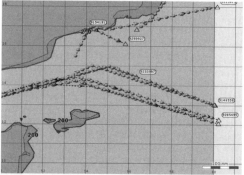

Fig. 3 – EU CDC anti-piracy tool

Based on the EU LRIT CDC and as requested by EU NAVFOR, the IMO has decided to extend the LRIT anti-piracy tool at international level in order to provide the navy forces in the area with a complete LRIT picture. Therefore the MSC .87 (May 2010) has agreed on the setting-up of a dedicated IMO LRIT anti-piracy facility which will provide LRIT reports of all ships transiting the area. This facility has become operational since July 2010 and all SOLAS CG can join the tool on a voluntary basis and provide the ship position of their ships to the navy forces patrolling the area.

This is one of the best positive examples on how the international cooperation can assist better implementation of the maritime security rules in high risk areas.

5 CONCLUSIONS

Worldwide sea traffic is increasing and security, safety and environmental risks are increasing too. Establishment of the LRIT system shall increase level of ships, coastal states and port states security and improve environmental protection, safety of navigation and efficiency of the search and rescue operations at high seas. It will increase the range of reporting requirements already imposed on ships engaged on international voyages by regulations either international (conventional) or regional and national introduced in a variety of places. International service providers should work on the basis of contracts (Public Service Agreements) signed between each one of them and IMSO or IMO. It is possible that particular flag states will reserve the right to approve service providers acceptable for their vessels.

It shall be stressed that LRIT system as described in this paper is technically operational at this stage.

Technologies are available to provide cost effective solution. Additionally according to the information presented by IMSO there is now about 45 000 ships which should participate in the LRIT system.

If all of them will send daily four reports for 20 cents, the total global cost will be around 13 140 000 USD per year. That is the reason that IMO and IMSO do not suspect any problems with finding the service providers. There are a number of parties with a legitimate interest in receiving LRIT data from ships:

– search and rescue, immigration, customs, quarantine and navigational services,
– security, environmental protection and Port State Control agencies,
– port authorities and ships' agents,

– commercial bodies (ships owners, cargo forwarders, charterers, etc); and
– fisheries management authorities.

Many different commercial and government owned and operated systems have been developed and introduced to cater for these interests. They vary in the type of technology used and costs of reporting a ship's position and related information. All existing conventional vessels engaged on voyages outside A1 sea areas are fitted and will be fitted with the terminals of the global satellite radio communication system Inmarsat-C for reception of Maritime Safety Information (MSI) and to meet other requirements of the Global Maritime Distress and Safety System (GMDSS). Those terminals can be used to transmit reports required by LRIT service without extra cost to the ship. Other ships may have to be fitted with additional equipment, but will be able to choose from a range of Communication System Providers.

Since 1st December 2004 the mandatory ship reporting system in the Great Barrier Reef and Torres Strait Vessel Traffic Service (REEFVTS) has been upgraded by introducing obligatory so called Pre-Entry Report and 15 minutes position updates transmitted via Inmarsat. It means REEFVTS creates already first in the world obligatory LRIT system for conventional vessels.

Experts unanimously stress that the only effective long-term piracy deterrent is a stable state. When Somalia was briefly under the control of the Islamic Courts Union in 2006, the piracy acts stopped completely.

Until recently, sovereignty has prevented outside states from targeting inland pirate infrastructure. A UN resolution passed on December 2, 2008, allows states to enter Somalia's territorial waters in pursuit of pirates, and another resolution passed on December 16, 2008, implicitly authorizes land pursuit.

REFERENCES:

[1] Wawruch R.: Conception of the Polish national ships monitoring system based on AIS technique. Zeszyty Naukowe Politechniki Śląskiej, seria Transport, z. 51, Gliwice, 2003, s. 543-549.
[2] Wawruch R.: Global ships monitoring system. Zeszyty Naukowe Politechniki Śląskiej, seria Transport, z. 59, 2005, s. 487-493. 66 R. Wawruch
[3] Wawruch R.: Global ships monitoring system – structure and principle of work. Monograph „Advances in Transport Systems Telematics", Section III „Systems in Maritime Transport", Chapter 4 "Global ships' monitoring system – structure and principle of work", Katowie, 2006, s. 307-312.
[4] MSC 81/25/Add.1, IMO, London, 2005.
[5] VTS 22/4/6, IALA, Saint Germain en Laye, 2005.
[6] Combating Maritime Piracy, Stephanie Hanson 2009.

20. A Decision Support Tool for VTS Centers to Detect Grounding Candidates

A. Mazaheri, F. Goerlandt, J. Montewka* & P. Kujala
Aalto University School of Engineering, Espoo, Finland
* Aalto University School of Engineering, Espoo, Finland, Maritime University of Szczecin, Poland

ABSTRACT: AIS (Automatic Identification System) data analysis is used to define ship domain for grounding scenarios. The domain has been divided into two areas as inner and outer domains. Inner domain has clear border, which is based on ship dynamic characteristics. Violation of inner domain makes the grounding accident unavoidable. Outer domain area is defined with AIS data analyzing. Outer domain shows the situation of own ship in compare with other similar ships that previously were in the same situation. The domain can be used as a decision support tool in VTS (Vessel Traffic Service) centers to detect grounding candidate vessels. In the case study presented in this paper, one type of ship, which is tanker, in a waterway to Sköldvik in the Gulf of Finland is taken into account.

1 BACKGROUND INFORMATION

1.1 *Ship Grounding*

Ship grounding accounts for about one-third of commercial ship accidents all over the world [1,2], and has the second rank in frequency, after ship-ship collision, in global perspective [3]. The consequences of ship grounding could be devastating for both humans and the environment. In less grave accidents, ship grounding might result in only minor damages to the hull; however, in more serious accidents, it might lead to the total loss of the vessel, oil spills and human casualties, in which the compensation would be either highly costly or even impossible. Therefore it would be wise to think about tools that can prevent ships to be involved in such accidents.

1.2 *Ship Domain*

One of the methods that have never been tried for grounding candidate detection is using the ship domain. The concept of ship domain has been first introduced by Fujii [4] in maritime transportation as an imaginary area around a ship, where the navigators try to keep it clear from other ships. Later on, Goodwin [5] redefined the concept as the effective area around a ship where navigators try to keep it clear from other ships and stationary objects. Since then, many other authors [6-16] have tried to define the size and shape of ship domain with different

methods. However, the main common issue in between all introduced domains is that all are suitable for ship-ship collision accidents, as the used methods are ruled by the nature of this type of ship accident. This fact is also recently highlighted by Wang [15,16]. Although some authors have mentioned their domains are suitable for grounding scenarios as well [5,6,13,14], the affecting factors that they have used to define the size and shape of the domain and also the application of the domains are more useful for ship-ship encounter situations. This is the main courage for the present research in defining a ship domain proper for ship grounding scenarios, in order to be used as a decision support tool in VTS (Vessel Traffic Service) centers to detect the ships that are grounding candidates.

2 SHIP DOMAIN FOR GROUNDING

Some factors that could affect the shape and size of a domain useful for grounding scenarios are ship main characteristics (length, breadth, draft, speed, and type), her maneuverability, navigator experience and his familiarity to the area, shape and depth of the waterway, engine and rudder characteristics, and weather condition; which some of them are not easy to consider and to model. In addition, the 3rd dimension (depth) is vital for defining the ship domain for grounding since the grounding is defined as the event that the bottom of a ship hits the seabed, in compare with stranding, which is defined as the event that a ship impacts the shore line and strands on shore [17]. Moreover, since normally ship has forward speed while goes aground, the domain for grounding could not be longitudinally symmetric. For the same reason lateral dimension of the domain should be always smaller than longitudinal dimension of the domain, when is defined for grounding and stranding cases.

One additional point about ship domain either for grounding or collision is that a domain should have two areas as they can be called inner and outer domains. *Inner domain* is the area, which is defined based on the dynamic of the ship. Because of the ship inertia, the ship's course cannot be altered in a moment. Inner domain defines the last/latest possible point/time that evading maneuver is possible for the ship by the most possible aggressive but safe maneuver, in order to avoid the accident. It means if the inner domain is violated by a shoal, even though the ship has not run aground yet, there is no way for her to survive an accident. *Outer domain*, on the other hand, can be defined as such that describes the area of different levels that mariners are advised to keep clear from any shoals or other stationary obstacles. Failing to do so, makes the vessel a grounding candidate with a certain degree. In contrary of the inner domain, the outer domain does not have clear border. Outer domain should be defined as such that if a ship does not do any evasive maneuver by certain time/distance, it is considered, by some degree, odd or unsafe for that particular ship with specific characteristics in specific situation and location.

It is worthwhile to mention, depends on the reason of the accident, ship grounding can be categorized into two major groups as powered and drift groundings. Nevertheless, drift grounding is a kind of accident that occurs as a consequence of an incident like engine or rudder failures, which makes the ship domain concept not applicable for this type of grounding.

3 METHODS TO DEFINE SHIP DOMAIN FOR GROUNDING

3.1 *Inner Domain*

The shape of the domain in this paper is taken as an imaginary half-elliptical prism. The ellipse is chosen to just explain the procedure of defining the size of the domain. To define a proper shape for the domain, in order to be rational for grounding accident analysis, more detailed data analysis and modeling are needed, which will be addressed in future studies.

The size of inner domain should be defined based on ship maneuverability, which is based on the dynamic of the ship. The length of the inner domain is defined to be equal to the summation of overall

length of the ship (LOA), influence region of ship-shore interaction (bank effect), and stopping distance or the advance in turning circle maneuver, whichever is shorter. To define the length of the inner domain in this paper, it is assumed that length of the advance in turning circle is smaller than the stopping distance, which is a valid assumption for ships moving with speed more than 12 kn [18]. The advance in turning circle in this paper is estimated with a quasi-linear modular hydrodynamic model of the vessel in-plane motion. For detail explanation of the used hydrodynamic model, the readers are referred to [19].

The width of the inner domain is taken equal to twice of the width of the influence region of bank effect. The influence region of bank effect (y_infls) in this paper is estimated based on a formula suggested by [20]. It should be mentioned that for defining the width of the inner domain it is assumed the ship does not comply with the given commands if she enters the influence region of bank effect. Therefore, controlling the ship will not be possible with ordinary skills, which makes the ship eventually hitting a channel bank. Although this assumption is not far from reality, it should be considered that some expert mariners might still be able to control the ship in that condition and therefore be able to survive from an accident. However, to define the inner domain, rare situations are neglected and it has been tried to define it as such to be suitable for majority of the cases.

The depth of the inner domain is taken equal to the maximum squat plus the draft of the vessel. The maximum squat in this paper is estimated based on a formula suggested by [21]. The schematic figure of the defined inner domain is shown in Figure 1.

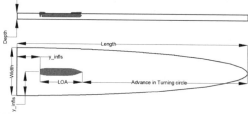

Figure 1: Three dimensions of the inner domain

3.2 *Outer Domain*

In this paper, outer domain is not defined by a unique imaginary shape; but as points in different waterway legs, in where the position and situation of the vessel is analyzed based on extensive AIS (Automatic Identification System) data analysis in respect to being a grounding candidate. The used algorithm is shown in Figure 2. The general idea is to choose a specific shoal/obstacle and analyze available AIS data transmitted by ships similar to the subject (own) ship, which have previously approached to the shoal, in order to find distribution for the longitudinal distance between ships and the shoal, in where ships start to turn to either evade the shoal or follow the fairway [Action Distance (AD), the point where it happens is named Action Point (AP)]. Thereafter, use the obtained probability density function (PDF) of AD to analyze the situation of subject ship in respect to the shoal, in regard to grounding accident. The PDF of action point will help the VTS operators to relate the present location of subject ship to the percentile of similar ships, which have chosen that specific location to start their maneuvers. In this regard, the appropriateness of the present location of the subject ship to start the turning maneuver can be judged by the safe maneuvers previously performed by ships similar to the subject ship. Similarity can be identified by indexes such as ship type, length, width, draught, speed, and even environmental conditions. The more indexes are defined, the more resembled cases can be retrieved and therefore the more reliable support for decision can be provided. However, more indexes need bigger and more complete databases to be used, in order to retrieve sufficient data for creating useful PDFs. Due to the scarce of data, the similarity in this paper is identified just by type and length of the ships.

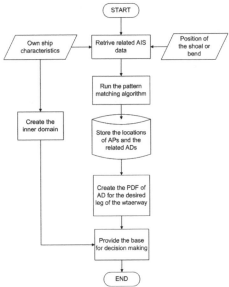

Figure 2: Algorithm to define outer domain

It should be borne in mind that because of the ship inertia, the ship's course cannot be altered in a moment; therefore it takes time between when the command is given to the controlling devices till when the command is started to be obeyed by the vessel, in where is defined to be Action Point. Nonetheless, this difference is neglected in this paper.

The action point detection process is based on a pattern matching algorithm shown in Figure 3. The pattern matching is based on course-over-ground (COG) of ships. The idea is to visualize COG of the ship in her path and then use the algorithm to detect the performed maneuvers based on the visualized COG. Here, visualizing means making the sequence of COGs smooth in order to not have any disruption in between. To explain visualizing and the algorithm, part of waypoints in a trajectory of a tanker in route from Sweden to Sköldvik in the Gulf of Finland (GOF) for year 2007 is shown in Figure 4-Left as an example. The history of COG of the shown trajectory of the tanker is shown in Figure 4 -Right.

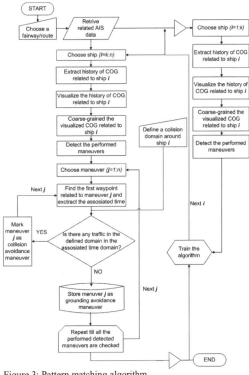

Figure 3: Pattern matching algorithm

The normal method being used for recording COG is to mark the heading to the North as $0°$, to the East as $90°$, to the South as $180°$, and to the West as $270°$ (turning clockwise). As a result, COG can never get negative values; and if, for instance, the ship is turning clockwise and COG value passes $359.99°$, the COG will be registered again as $0°$. Therefore, if the graph of the history of COG be drawn, there might be some jumps in the graph (Fig. 4). To remove the disruptions and making the sequence of COGs smooth (visualizing), the COGs are transferred to another discipline that is shown in Figure 5. In the new discipline COG can get negative values as well as values more than $360°$. The history of

COG after visualizing is shown in Figure 4, which shows the jumps are disappeared. The visualized COGs of ships navigating in a fairway are somehow unique for the fairway, and can act as fingerprint of the fairway, which the pattern matching algorithm can recognize. By knowing the position of turns/shoals in a fairway and having the visualized COGs of the ships navigating in the same fairway, the evasive maneuvers that have been done to follow the turn/avoid the shoal can be identified. The starting point of the associated maneuver is stored as AP and the shortest distance between AP and the shoal is reported as AD. It should be added that for decreasing the margin of error for pattern matching, the visualized COGs are coarse-grained in order to remove the small changes in COG, which are normally appears due to course adjustment. Moreover, to minimize the possibility of choosing a collision avoidance maneuver, the presence of ship traffic in instance time domain in an area around the vessel, which is defined by the domain proposed by [15] for collision scenarios, is also investigated and taken into account.

bility of data, and also because of the importance of grounding accident in the area. The studied area is a waterway in GOF with approximate length of 40 km, in where the ships have to navigate in between shoals in order to reach to Sköldvik. The waterway is located in a rectangle which end points of one of its hypotenuses have positions of 60.0° N 025.4°, E, and 60.4° N, 025.7° E in WGS-84 reference system (Fig. 6-Left). The majority of the traffic in this area belongs to tanker traffic. Therefore, the other types of ships are eliminated from the analyzed data due to data scarce. In total 850 tankers navigated in that area in 2007 with the shortest length of 75 m and the longest length of 265 m. The AIS data analysis is done with Mathwork's MATLAB. Thus, for the sake of coding, the shoals in the area are defined as polygons. In total, five shoals in the area are defined and taken into account for data analysis. The shoals and vertices of the polygons are shown in Figure 6-Right.

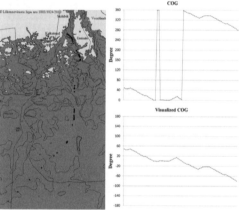

Figure 4: Left: Part of a trajectory of a tanker in route from Sweden to Sköldvik in GOF for year 2007, Right: COG and visualized COG of the trajectory of the tanker

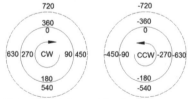

Figure 5: Discipline used for visualizing the history of COG

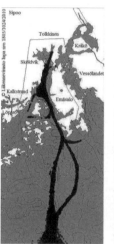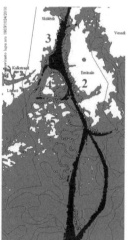

Figure 6: Left: The waterway to Sköldvik in the Gulf of Finland with the traffic in 2007- Right: The same waterway with the analyzed shoals as polygons. The vertices of the polygon shoals are shown in dots.

4 CASE STUDY AND RESULTS

The analyzed AIS data in this paper is for the year 2007 of ship traffic in the Gulf of Finland, which was gathered by the Finnish Transport Agency. The Gulf of Finland is used for the study due to availa-

To define the domain in order to be used for VTS operators, PDF of AD for the ships in each leg of a waterway should be extracted. Based on the extracted distributions, inner domain, and speed of the vessel, the VTS operator can have a good analysis of the present position of the vessel. By way of illustration, it is assumed that the subject ship is a most common tanker for this harbor, with the dimensions of L=145 m, B=17 m, T=10 m navigating in the studied waterway with speed of 15 kn. Using ship type and ship length as indexes, the related PDF for AD can be extracted from the database. The PDFs for shoals 1 and 2 are shown in Figure 7 as examples. With the help of the extracted PDFs, the percentile of the similar ships that have started to turn by specific point in the same leg of the waterway can be estimated. In addition, the defined inner domain gives the remained time to go aground on approaching shoals. The inner domain for the studied tanker is estimated based on the advance of turning circle in maximum rudder angle, which is assumed to be 35°. Example of analysis of nine positions of the chosen tanker in the studied area (Fig. 8) is shown in Table 1 as a way of illustration.

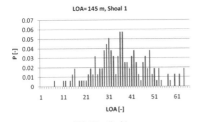

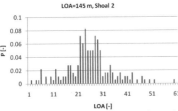

Figure 7: PDF of Action Distances for tankers with LOA of 145 m approaching shoals 1 and 2

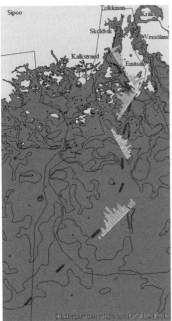

Figure 8: The subject ship (L=145 m, B=17 m, V=15 kn) in a way to Sköldvik shown with her inner domain. The dark areas are the inner domains. The tanker is seen as a small black dot in this scale

It can be seen in Table 1, wherever the outer domain shows that the majority of the similar ships, previously navigated in the same waterway leg, had started their turning maneuver in that specific position to either evade the shoal or follow the waterway, the inner domain shows less available time for maneuvering in order to avoid grounding. Information as such will help the VTS operators to detect those ships that their remaining chance to survive from a grounding accident are getting less and less, with the aim of marking them as the ship that her actions should be monitored more closely. Later one, the VTS operator may decide to contact the ship to find if the officer on watch is aware of the situation. In this way, the VTS operators are capable of being more proactive.

5 DISCUSSION AND CONCLUSION

A new approach to define ship domain for grounding scenarios based on AIS data analyzing and ship maneuverability is presented in this paper. The introduced domain is suggested to be used as decision supports tool in VTS centers. It is shown the introduced domain is capable of providing useful information, like remaining time to point of no-return and going aground, based on the vessel maneuverability. In addition, the proposed method is able to provide the ground for judging the safeness/oddness of the performing maneuvers. Since the method uses previously performed maneuvers to analyze the current maneuvering action, it can be argued the method is providing expert opinions as a support for decision making process.

The turning circle and stopping distance are used in the definition process of the inner domain for grounding in this paper. Since those concepts are unique for every single vessel in unique conditions, this method neutralizes the effects of type and number of controlling devices in hand. Nonetheless, it makes hard to estimate the area of inner domain precisely, as the available hydrodynamic models for predicting the ship motion are not completely flawless. However, the quasi-linear modular hydrodynamic model used in this paper can predict the turning circle of vessels precisely enough for the scope of this paper [19]. In addition, using turning circle to define inner domain area limits the usability of the suggested domain to when all reserved maneuverability of the ship is available, which means when the vessel is moving straight. The maneuvering task is somehow different while the ship is in turning process, as she does not have all the reserved maneuverability in hand. Due to this fact, grounding ship domain in complex turns might be different than what has been introduced in this paper.

Table 1: Situation analysis of the subject tanker in nine positions shown in Figure 8

Position	COG [deg]	Percentile of similar ships that have started to turn	Time to breach inner domain, maintaining COG and speed [min]	Time to ground on the approaching shoal, maintaining COG and speed [min]
1	48	0%	54	56
2	45	0%	36	38
3	48	5%	15	17
4	18	38%	10	12
5	0	6%	15	18
6	12	4%	15	17
7	15	83%	6	8
8	331	75%	7	9
9	338	83%	6	8

* Position numbering is started from left-down corner of the Figure 8. The first position in left-down corner is position 1, next position is 2 and so on.

The analyzed AIS data used for defining outer domain in this paper are indexed based on ship type, ship length and the location. This has been done due to the scarce of the data. By increasing the size of the used database and also using data about weather and sea conditions, the indexes can be expanded to other characteristics of the vessel and also to environmental conditions, in order to provide more reliable supports for decision making process. Moreover, the analyzed data are limited to year 2007. Analyzing more data from other years will help the used algorithm to be more precise in providing the grounds for decision making. In addition, the algorithm can be made smarter if a learning loop be added, in order to teach the algorithm by new performing maneuvers.

The introduced domain is proposed as a decision support tools for VTS centers. Nevertheless, it is possible that the introduced domain be used as a decision support tools onboard the vessels, in order to provide expert opinions for officer on watch to perform maneuvers.

AKNOWLEDGMENT

The authors wish to thank the financial contribution of the European Union and the city of Kotka. This research is carried out within SAFGOF project and in association with Kotka Maritime Research Center.

REFERENCES

1. Kite-Powell, H.L., et al., *Investigation of Potential Risk Factors for Groundings of Commercial Vessels in U.S. Ports.* International Journal of Offshore and Polar Engineering, 1999. 9(1): p. 16-21.
2. Jebsen, J.J. & V.C. Papakonstantinou, *Evaluation of the Physical Risk of Ship Grounding*, in *Department of Ocean Engineering.* 1997, Massachusetts Institute of Technology. p. 239.
3. Samuelides, M.S., N.P. Ventikos, & I.C. Gemelos, *Survey on grounding incidents: Statistical analysis and risk assessment.* Ships and Offshore Structures, 2009. 4(1): p. 55-68.
4. Fujii, Y. & K. Tanaka, *Traffic Capacity.* The Journal of Navigation, 1971. 4: p. 543-552.
5. Goodwin, E.M., *A statistical study of ship domains.* The Journal of Navigation, 1975. 28(3): p. 328-344.
6. Davis, P.V., M.J. Dove, & C.T. Stockel, *A computer simulation of marine traffic using domains and arenas.* The Journal of Navigation, 1980. 33: p. 215-222.
7. Coldwell, T.G., *Marine Traffic Behaviour in Restricted Waters.* The Journal of Navigation, 1983. 36: p. 430-444.
8. Zhao, J., Z. Wu, and F. Wang, *Comments on Ship Domains.* The Journal of Navigation, 1993. 46(3): p. 422-436.
9. Lisowski, J., A. Rak, & W. Czechowicz, *Neural netwrok classifier for ship domain assessment.* Mathematics and Computers in Simulation, 2000(51): p. 399-406.
10. Zhu, X., H. Xu, & J. Lin, *Domain and its model based on neural networks.* The Journal of Navigation, 2001. 54(1): p. 97-103.
11. Pietrzykowski, Z. & J. Uriasz, *Methods and criteria of navigational situation assessment in an open sea area*, in *4th International conference on computer and IT applications in the maritime industries, Compit'05.* 2005: Hamburg, Germany.
12. Pietrzykowski, Z. & J. Uriasz, *Ship domain in navigational situation assessment in an open sea area*, in *5th International conference on computer and IT applications in the maritime industries, Compit'06.* 2006: Oud Poelgeest, Leiden, Netherlands.
13. Pietrzykowski, Z., *Ship's Fuzzy Domain - a Criterion for Navigational Safety in Narrow Fairways.* The Journal of Navigation, 2008. 61: p. 499-514.
14. Pietrzykowski, Z. & J. Uriasz, *The Ship Domain - A Criterion of Navigational Safety Assessment in an Open Sea Area.* The Journal of Navigation, 2009. 62(1): p. 93-108.
15. Wang, N., *An intelligent spatial collision risk based on the quaternion ship domain.* The Journal of Navigation, 2010. 63: p. 733-749.
16. Wang, N., et al., *A unified analytical framework for ship domains.* The Journal of Navigation, 2009. 62: p. 643-655.
17. Mazaheri, A., *Probabilistic Modeling of Ship Grounding - A review of the literature*, in *TKK-AM-10.* 2009, Helsinki University of Technology: Espoo. p. 73.
18. Mandel, P., *Ship Maneuvering and Control*, in *Principles of Naval Architecture*, J.P. Comstock, Editor. 1967: New York. p. 463-606.
19. Montewka, J., et al., *Probability modelling of vessel collisions.* Reliability Engineering and System Safety, 2010. 95: p. 573-589.
20. Lataire, E., et al., *Navigation in confined waters: Influence of bank characteristics on ship-bank interaction*, in *International Conference on Marine Research and Transportation.* 2007: Ischia, Italy. p. 135-143.
21. Derrett, D.R., *Ship Squat*, in *Ship Stability for Masters and Mates*, C.B. Barrass, Editor. 1999.

21. On the Development of an Anchor Watch Supporting System for Small Merchant Ships

H. Yabuki & T. Takemoto
Tokyo University of Marine Science and Technology, Tokyo, Japan

K. Yamashita & S. Saitoh
National Institute for Sea Training, Yokohama, Japan

ABSTRACT: This paper describes the results of a study that aimed at developing an effective anchor watch supporting system to prevent dragging anchor accidents of small domestic merchant ships. The authors performed an experimental study using a training ship in order to investigate the characteristics of the hull movement of a ship lying at single anchor, the cable tension caused by the above movement and etc. Based on the results of the study, the authors propose a standard procedure for safe anchor watch and a new anchor watch supporting system using a PC, a DGPS and an anemometer.

1 INTRODUCTION

In the Japanese coastal waters, many disasters involving small domestic merchant ships dragging anchor are reported every year. Domestic merchant ships often anchor temporarily when waiting for their berths or avoiding storms. Accidents by dragging anchor often occur during the above temporary anchoring and it is reported that no anchor watch is provided in most of the accidents because domestic merchant ships are operated by a limited number of crew members.

One way to prevent accidents caused by dragging anchor is to develop an anchor watch supporting system that will monitor the anchoring condition and detect a risk of dragging anchor in advance.

From the above points of view, the authors conducted the full-scale experiments in order to investigate the following characteristics of a ship at anchor; the eight-figure horsing movement of the hull (horsing) when lying at single anchor, the cable tension caused by the horsing, the hull movement at dragging anchor, the effect of a secondary short-scope anchor to reduce the horsing.

The test ship was a 5,800 G.T. training ship and her principal particulars are shown in Figure 1.

Based on the results of the above experiments, the authors proposed a method to establish the standards of safe anchoring and anchor watch to prevent a dragging anchor accident. A new anchor watch supporting system with a function to detect a risk of dragging anchor is developed for small domestic merchant ships.

2 STANDRDS OF THE ANCHORING

2.1 *Horsing movement lying at single anchor*

The eight-figure horsing movement of the hull (horsing) when lying at single anchor and the cable tension caused by the horsing were measured using the test ship (Saitoh 1986).

Figure 2 shows an example of the time history of the cable tension during horsing when lying at single anchor with 7 shackles of her cable. The depth of sea water was 27m and the wind velocity was 20 m/s in average. The strong cable tension (shock load that acted on her cable) was measured 4 times during one cycle of horsing.

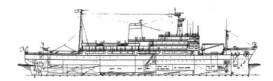

Hull		Anchor	
G.T. (ton)	5886.0	JNR type	3.455 ton x 2
Loa (m)	124.8	Cable (56φ)	250 m x 2
Lpp (m)	115.0		
B (m)	17.0		
D (m)	10.5		

Figure 1. Principal particulars of the test ship

The relationship between the measured shock load and the wind velocity is shown in Figure 3. The values of the cable tension in this figure are the average of the above shock loads that appeared 4 times during one cycle of horsing.

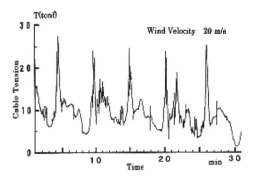

Figure 2. Time history of the cable tension during horsing

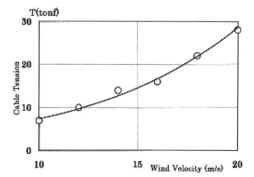

Figure 3. Relationship between the cable tension and the wind velocity

The horsing movement was observed when the wind velocity exceeds 10 m/s and the shock load seems to increase significantly in the range of wind velocities of 16 m/s or more. It is well known that this shock load is one of the main causes of dragging anchor. When the shock load is greater than the holding power of an anchor, it pulls the anchor and the flukes start overturning. The holding power of an anchor is reduced significantly due to the overturning of flukes. An overturned anchor can not bite into the seabed firmly and starts sliding over the bottom.

2.2 Effect of the secondary short-scope anchor

There is a possibility to reduce the shock load by controlling the degree of horsing, and some horsing control methods have been proposed. The use of a secondary short-scope anchor is one of the effective and easy countermeasures to reduce horsing. This method utilizes a dragging resistance of the secondary short-scope anchor and the cable length of it is recommended to be 1.25 to 1.5 times the depth.

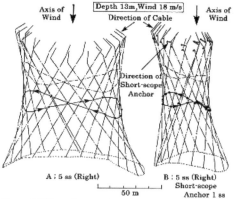

Figure 4. Effect of a secondary short-scope anchor to reduce horsing

Table 1. Example of the Standards of anchor watch (Depth; 15m)

wind (m/s)	Counter measure
15 or less	5 ss
15 – 17	6 ss
17 – 20	7 ss, short-scope anchor
20 - 22	8 ss, short-scope anchor Office's anchor watch
22 or more	S/B Eng. & Rudder

Figure 4-A shows the trajectory of the test ship while horsing on a single anchor with 5 shackles of her starboard cable. The water depth was 13 m and the wind velocity was 18 m/s in average. The test ship sheers violently back and forth across the wind. Figure 4-B shows the hull movement of horsing with 5 shackles of riding cable and one shackle of the secondary short-scope anchor under the same conditions. In this case, the lateral movement range of the center of gravity with a secondary short-scope anchor is 50 % smaller than that without a short-scope anchor. As these experiment results agree with those of model tests qualitatively, it seems that the use of a secondary short-scope anchor is very effective and useful to reduce horsing.

2.3 Standards of the anchoring and anchor watch

Table 1 shows the standard procedure of the anchoring and anchor watch for the test ship when anchored at 15 m depth of water. Usually, the cable length that should be veered out at the first stage of anchoring is calculated by the following formula; $(3D + 90)$ m, D means the depth. This empirical formula is widely used among the Japanese seafarers. After anchoring, the test ship veers out her cable and drops a secondary short-scope anchor according to the increase of wind velocity. When wind exceeds the designated velocity, an officer's anchor watch is started for earlier detection of the risk of dragging anchor, and her main engine, rudder and other necessary machinery are prepared.

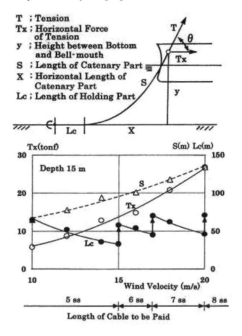

Figure 5. Estimated length of holding part during anchor watch by the standard procedure

Figure 5 shows the estimated length of cable that remains on the sea bed (L_c ; holding part length of the cable) when the test ship performs her anchor watch in accordance with the above standard procedure. The holding part length (L_c) is estimated using the following equations and the horizontal force of shock load (T) shown in Figure 3.

$$
\left.
\begin{aligned}
L_c &= L - \sqrt{y\left(y + \frac{2T_x}{w'_c}\right)} \\
T_x &= T - w'_c \cdot y
\end{aligned}
\right\}
\tag{1}
$$

where L = total scope; and w'_c = weight of chain per unit in the sea water.

A certain length of the holding part should be kept for the safe anchoring because it acts as a spring in preventing the anchor from being jerked when the ship is yawing from side to side. In the case of the test ship, her cable is veered out in accordance with the increase of wind velocity in order to keep the holding part length at least two shackles. To include the above method in the standard procedure of anchor watch is considered to be useful and helpful for small domestic merchant ships.

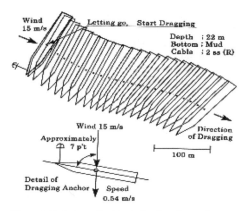

Figure 6. Trajectory of a ship during dragging anchor

3 DEVELOPMENT OF AN ANCHOR WATCH SUPPORTING SYSTEM

3.1 Detection of a risk of dragging anchor

Figure 6 shows the trajectory when the test ship drags her anchor under 15 m/s of wind. The hull is drifted at a very slow speed of 0.54 m/s to the leeward by the beam wind. Her heading is about 7 points to the left of the wind axis during dragging anchor. As the above experiment results agree with the simulation results (Inoue 1988), we can conclude, when the regular horsing movement is stopped and ship's weather side becomes fixed, that the ship is likely to be dragging anchor.

Once an anchor starts to drag, it is difficult to stop. Therefore, it is important to detect the possible danger of dragging anchor at an earlier stage in order to take counter measures to prevent it in advance.

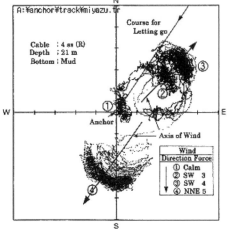

Figure 7. Trajectory of horsing lying at single anchor (96 hours)

Figure 7 shows the 96 hours trajectory of the test ship lying at single anchor with 5 shackles of her cable. The location of horsing is moved in accordance with the change of wind direction and its force. The center of the horsing is not located at the anchor position but around the point at the end of the Catenary part of the cable that is touching the sea bed. As described in 2.3, this is due to a certain length of the cable is always kept on the sea bed during anchoring. Consequently, there is a possibility to judge the existence of the risk of dragging anchor when the length of Catenary part is equal to that of riding cable and the holding part is missing.

The authors propose the method to detect the risk of dragging anchor that compares a horizontal distance between anchor and the bell-mouth (d) and horizontal length of the Catenary part (X_{max}) when its length is equal to the riding cable length as shown in Figure 8. The distance "d" is monitored using the DGPS. The diagram in the Figure 8 shows the time history of the distance "d" when the test ship was anchored in 18 m depth of the water under 20 m/s wind with 8 shackles of riding cable and a secondary short-scope anchor. The time history reflects the back and forth movement of the test ship that is induced by the horsing.

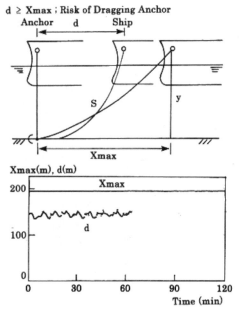

Figure 8. Method to detect the risk of dragging anchor and monitoring result of the distance "d"

The possible danger of dragging anchor can be judged using the following observation results.
1 Existence of a strong wind
2 The direction of anchor almost agrees with the wind axis.
3 The trend of "d" approaches toward the threshold "X_{max}".

3.2 Anchor watch supporting system

As domestic merchant ships are operated by a limited number of crew, the following functions are desired for their anchor watch supporting system.
1 Detecting function of dragging anchor and possible danger of dragging anchor.
2 Monitoring function of wind, horsing movement, hull posture against the wind axis, location of an anchor and etc.
3 Anchor watch supporting function
4 Alarm function

The authors propose a simple and user friendly anchor watch supporting system with the above functions. This system consists of a PC, DGPS, Gyro Compass and Anemometer, which are common navigational equipment onboard domestic merchant ships.

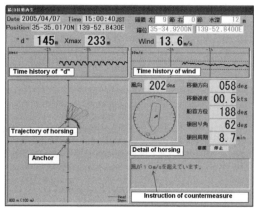

Figure 9. Display of the anchor watch supporting system

Figure 9 shows the display of the system that was developed for the test ship and the following data can be monitored; horsing trajectory, heading, moving direction and speed of hull, horsing angle, location of anchor, hull posture against the wind axis, wind direction and its velocity. The possible danger of dragging anchor can be detected by comparing the displayed time history of "d" and "X_{max}". When the wind velocity exceeds the threshold that is designated in the standard procedure of anchor watch shown in the Table 1, the necessary countermeasure in accordance with the procedure is displayed with alarm.

4 CONCLUSION

The authors performed an experimental study in order to develop a method to establish the standard procedure of safe anchoring and an anchor watch supporting system for small domestic merchant ships. Results obtained in this study are summarized as follows.

1 A ship at single anchor starts horsing at the wind velocity of 10 m/s and the shock load caused by horsing acts on her cable. Ship sheers violently back and forth across the wind when it is at 16 m/s or more and the shock load increases remarkably.
2 The secondary short-scope anchor is very effective to reduce horsing in stormy weather.
3 For the prevention of disaster caused by dragging anchor, it is important to establish the standard procedure of anchoring and anchor watch. The proposed method for establishing the above procedure is considered to be effective to prevent dragging anchor.

4 The developed anchor watch supporting system with a function to detect a risk of dragging anchor in advance is useful and effective to prevent the dragging anchor accident.

REFERENCES

Saitoh, S. et al. 1986. A study on anchoring in stormy weather –On the measurement of ship's cable tension at anchor-, *Journal of Japan institute of navigation, Vol.* 74; 9-18 (in Japanese)
Inoue, K. 1988. Countermeasures to assure the safety of the outside-harbour-refuge, *Journal of Japan institute of navigation, Vol.* 78; 129-138 (in Japanese)

22. Integrated Vessel Traffic Management System (VTMIS) for Port Security in Malaysia

N.A. Osnin, A.H.Saharuddin & Z. Teh
Faculty of Maritime Studies and Marine Science, Universiti Malaysia Terengganu

ABSTRACT: Vessel Traffic System (VTS) today are no longer a privilege reserved for major ports, but are becoming increasingly significant for all ports to manage traffic safely and efficiently, coordinate emergency response effectively and continuously vigilance of security threats. Integrated Vessel Traffic Management and Information System (VTMIS) focuses on integrating traffic management, providing situation awareness, maintaining security surveillance, coordinating emergency responses and facilitating data interchanges. This system also provides common-operational-picture visualisation for ports and its 'allied services'. With the advent of maritime security regimes such as the International Ship and Port Facility Security (ISPS) Code, this solution is needed to fulfil the continually increasing security standards. This paper highlights the developments of VTMIS in Malaysian ports.

1 INTRODUCTION

Seaport security has become a critical element of supply chain economics and a vital concern for world shipping. Every seaport and its waterway present unique safety and security challenges and the vulnerability issue usually requires consideration of large-scale, multitier complex systems. Seaport and its waterway must be protected from illegal activities and environmental hazards. Cargo and ships need to get to their intended destination safely, on schedule and need protection from any security threat. Managing safety and security in port waterway area requires an affordable and versatile system, technology and components that can deliver high-performance and high reliability. As a result, Vessel Traffic Management and Information System (VTMIS) are seen as the best mechanism to be integrated in managing security in the Malaysian ports and waterways.

1.1 *Why VTS ?*

Vessel Traffic Services (VTS) evolved as a response to the increased complexity of modern shipping, the diversity and potential danger resulting from their cargoes and the need to prevent congestion by maintaining a safe traffic flow. VTS are designed to provide support to the mariner in busy waterways where risks are deemed at their greatest. VTS systems are now being recognized as a significant contributor to enhanced port and waterway safety and security.

1.2 *What is VTS ?*

A VTS is a service implemented by a competent authority, designed to improve the safety of vessels traffic and to protect the environment. The service should have the capability to interact with the traffic and respond to traffic situations developing in the VTS area.

The International Maritime Organisation (IMO) Assembly Resolution A.857(20), Guidelines for Vessel Traffic Services, established that the following tasks should be performed by a VTS:

A VTS should at all times be capable of generating a comprehensive overview of the traffic in its service area combined with all traffic influencing factors. The VTS should be able to compile the traffic image, which is the basis for the VTS capability to respond to traffic situations developing in the VTS area. Traffic image allows the VTS operator to evaluate situations and make decisions accordingly. Data should be collected to compile the traffic image. This includes:

1 Data on the fairway situation, such as meteorological and hydrological conditions and the operational status of aids to navigations;
2 Data on the traffic situation, such as vessel oppositions, movements, identities and intentions with respect to manoeuvres, destination and routing.

Data of vessels in accordance with the requirements of ship reporting and, if necessary, any additional data required for the effective operations of VTS.

1.3 The Development of VTS

Traditionally, the master of a ship is responsible for a ship's course and speed, assisted by a pilot where necessary. Ships approaching a port would announce their arrival using flag signals. With the development of radio in the late 19th century, radio contact became more important. But the development of radar during World War Two made it possible to accurately monitor and track shipping traffic.

The world's first harbour surveillance radar was inaugurated in Liverpool, England in July 1948 and in March 1950; a radar surveillance system was established at Long Beach, California. The ability of the coastal authority to keep track of shipping traffic by radar, combined with the facility to transmit message concerning navigation to those ships by radio, therefore constitute the first formal VTS systems.

The value of VTS in navigation safety was first recognised by IMO in resolution A.158 Recommendation of Port Advisory Systems adopted in 1968, but as technology advances and the equipment to track and monitor shipping traffic became more sophisticated, it was clear that new guidelines were needed on standardising procedures in setting up VTS.

As a result, in 1985, IMO adopted resolution A.578(14) Guidelines for Vessel Traffic Services, which said that VTS was particularly appropriate in the approaches and access channels of a port and in areas having high traffic density, movements of noxious or dangerous cargoes, navigational difficulties, narrow channels, or environment sensitivity. The Guidelines also made clear decisions concerning effective navigation and manoeuvring of the vessel remained with the ship's master.

2 VTMIS

2.1 The emergence of VTMIS

VTS when includes traffic and resources planning tools it is called VTMIS. Resource planning tools include:
– Tracking system
– Management information system – store and retrieve vessel and journey for the VTS
– Recording system
– Control and monitoring

Vessel Traffic Management and Information System aimed at improving the maritime transport system by implementing efficient traffic management, from which a wide variety of actors in maritime domain could benefit. When talking about improvements in the maritime transport system we have to consider the following objectives, which are generally accepted and expressly mentioned.

These objectives are to enhance the traffic safety, efficiency of the transport system and protection of the maritime environment. A wide range of actors are involved, among them the ship's crew, ship owners, ports, maritime enforcement agencies, search and rescue organisations and pollution control authorities. Only if these actors cooperate in an optimum manner and according to previously agreed interfaces and procedures the transport system will be able to meet the set objectives.

In the middle of the 80s, VTS were only installed in a few major ports having difficult navigational conditions or very dense traffic in their area. The integration of Electronic Data Interchanges (EDI) into VTS in order to form a VTMIS provides all relevant data regarding the vessel and its cargo needed for regular VTS management purposes, for 'value added services' and, particularly, in emergency situations. This denotes the exchanges or information with a wider scope of maritime actors than in standard VTS. The information exchanges are intended not only with the so-called 'allied services', i.e. search and rescue or pollution control, but also 'value added services', i.e. services to facilitate the logistics chain.

VTMIS are services, not systems. VTMIS is a concept, a kind of umbrella, for all activities to improve vessel traffic information. VTMIS delivers a traffic image to be used by authorities, ports and companies involved in vessels and cargoes. VTMIS can be used for vessel traffic management, port resources management, fleet management and cargo flow management. VTMIS requires an authority or service provider as a driving force of catalyst and 'win-win' objectives for all parties concerned. VTMIS can best be developed 'bottom up' but with the need of others in mind. In ports, VTMIS offers a considerable increase in operations efficiency and probably a fast return on investment, as several organisations benefit from the VTMIS-type of information exchange.

The principal development in VTS in recent years has been change from 'traffic monitoring' to 'traffic management' by the introduction and interconnection of databases and experts systems. They access static and dynamic information concerning ships, their cargo, and port services requirements. Together with an automatic update of tracking information, the VTMIS provides a powerful tool for the programming of traffic within the surveillance area. Operator can associate tracked targets with vessels registered in the database, which makes the data readily available and allows the system to automatically provide pertinent voyage information to other port service providers.
– Static vessel data (physical characteristics, owner, agents)
– Dynamic vessel data (cargo, previous voyage movements, planned movements)
– Records of incidents (planned/unplanned events, accidents)

- Resource allocation (pilots, berths)
- Accounting information (time in port, piloting charges, cargo loaded)

2.2 *Integration is KEY*

An integrated VTMIS can include many target sensors. When multiple sources of target information are involved, the data from individual radars and transponders in integrated so that a single target is presented to the operator. This also enables the system to offer a totally integrated information service that can be linked to all other port operational areas and functions, such as the port management system or pilot allocation system. In this way, port operations can be coordinated for the safe and efficient movement of vessels, and to achieve the optimum use of berth space and other port facilities, thus minimising delays to ships using the port.

Today's vessel traffic management equipments is very sophisticated with coming of new technology, the integrated system may comprise several or all of the following subsystems:
- Radars;
- Automatic Identification System (AIS) base stations;
- CCTV camera system (possibly with thermal imaging and infrared capability);
- VHF radio stations;
- VHF and EPIRB/MOB direction finders;
- Meteorological sensors (anemometer, thermometer and barometer);
- Hydrological sensors (tide gauge and current meter);
- Operator Display Unit (ODU) with ECDIS and ENC overlays;
- Recording and playback system;
- Simulation training system;
- Port management system;
- Emergency response system (oil spill management system);
- Geographical Information System (GIS);
- Statistic and report generation system;
- Web-based integration with other 'allied services'; and
- Integration with Pilot Carry Aboard Package (PCAP).

2.3 *VTMIS and the ISPS Code*

- ISPS Code requirements for ports naturally merge with general concept of traffic monitoring and VTS technology. Products and solutions, successfully used in VTS, perfectly suit ISPS requirements, which may be condensed in the following basic tasks:
- Collection of data providing sufficient awareness about traffic situations in the area, and security zones monitoring;

- Alarm generation according to a preset criteria, individually configured for each on/offshore area according to a defined security level;
- Data recording for later playback and specific situation debriefing;
- The capability of advanced functionality in case of the need to switch to a higher security level.

Most ports are identified by security assessments as being areas of high risk for terrorist attacks. Thus, part of a port security system (except for land based CCTV systems, fencing, security gates, etc) must provide independent traffic monitoring. At lower security levels, an ISPS system may be unmanned and left for automated performance with data recording. At any time a watch officer may take over the system control, transforming it into Security Coordination Centre.

Despite the mandatory implementation of AIS on most commercial fleets, radar sensor remains the main source of traffic situation data in an ISPS system. Radar tracking is generally not accurate as AIS position report, but radar return (echo) comes from almost any object, especially those which don't want to be seen. AIS recipients still depend on the quality of the onboard AIS installation, adequacy of data entered by shipmasters, and whether AIS onboard transponders are switched on or not.

3 MALAYSIAN PORTS

Ports are at the forefront of trade facilitators propelling Malaysia's economic growth and linkage to the world. It plays a strategic role that is crucial to the nation's economic wellbeing. It is estimated that 90% of Malaysian trade are seaborne and being one of the world's top 20 largest trading nations, the nation's dependence on its ports cannot be overstated. It is therefore crucial that the stakeholders involved in ports seriously focus their efforts to promoting and enhancing the competitiveness of local ports. While concrete actions have been and are being taken by various parties i.e. The Government, port operators, port authorities, etc. towards meeting this objective, they should not be complacent and should strive to further enhance the competitiveness of Malaysian ports.

3.1 *Shipping Traffic*

Shipping traffic ranging from small supply barges to vessel associated with the fishing trade to cargo vessels transiting have been growing steadily. These can be divided into two broad categories. The first results from the substantial and growing maritime trade. The vessels involved in this trade are a mix of large commercial ships classed as containerships, bulk carriers, car carriers, tankers and other commercial vessels.

The second category of vessels operating in Malaysian ports and waterways include local fishing vessels, supply, work and service vessels for the offshore industries and recreation crafts. Vessels in this category are smaller, carry less fuel and cargo and are typically on shorter voyages.

Inevitably, as the population of shipping traffic grows so will the likelihood of accidents and environmental catastrophe. Port areas are often in built-up areas and close to housing and other communal facilities i.e. adjacent to fisheries, wildlife and recreational areas. Thus comprehensive and efficient maritime traffic management systems become necessary.

This necessity is confirmed by accidents and disasters that have happened in Malaysian ports and waterways. Examples include the accident in Penang Port between two high speed ferries or the collision between a pilot boat and an unlit barge within an anchorage area. These could have been avoided by an effective VTMIS.

3.2 Role of Port Authorities in Malaysian Ports

Port Authorities have clear legal duties and powers to ensure the safety of their jurisdiction and therefore must come up with a proactive approach to safety management. They are also accountable for setting standards, allocating resources to safety and adopting effective systems. However, it is also acknowledged that commercial pressures and funding obligations does have an impact on these issues.

Financial and administrative assistance must be provided to initiatives promoting a culture of excellence in port management and operations, specifically in the management of maritime traffic within their jurisdiction.

3.3 VTS in Malaysian Ports

The Port Klang VTMS was developed in conjunction with the Malaysian Sea Surveillance System (MSSS) in 1997 and was commissioned in 2000. Maritime traffic management services in other ports such as Johor Port, Penang Port and Kuantan Port are still at the minimum level and are operated by their respective Licensed Port Operators which mostly deals only with port operations i.e. deploying of pilots, tugs, pilot boats, etc.

Bintulu Port has taken proactive action and appointed Kongsberg Norcontrol IT to upgrade its VTMS. This began in 2008 and is ongoing with the contract due in 2013. The Bintulu Port system solution includes C-Scope Extractor and Tracker represents superb value as it offers a wealth of operator functionality in addition to improving radar performance on minor maritime traffic within the port areas.

Much needs to be done by other Port Authorities in Malaysia to enhance the maritime traffic services in their ports. A good maritime traffic management in ports and waterways will improve efficiency and reduce the risk of marine accidents by providing timely, accurate and relevant information. However, as each individual port is of different nature and design, each must implement a system that meets its individual needs.

Stake holders must identify and share best practices. It should adopt the latest technologies, methods, systems and processes. A properly staffed, managed and equipped VTMIS is a valuable and versatile tool that can be integrated with other services and functions including pilotage, aids to navigation, search and rescue, law enforcement as well as customs and immigration.

4 CONCLUSION

Maritime traffic volumes in the Malaysian port waterways continue to increase and given trends with growth of regional economies and seaborne trade, shipping traffic growth will continue into the future. Analyzing these trends is an important contribution to understanding regional maritime safety requirements and a key factor in decision making for those responsible for maintaining security and safety in Malaysian Ports.

The main safety concern related to the increasing maritime traffic is the increased risk of collisions between different types of vessels, in particular passenger vessels, recreational and traditional traffic intersecting busy tanker routes. This necessitated an investment in risk control mechanisms such as a full fledged VTMIS.

A maritime traffic management system can have maximum benefit if it is established according to the actual and identified needs of the port, sea or river areas. With a good VTMS, the ports not only contribute to safety but also offer value to its customers by improving quality of service.

REFERENCES

Barco Traffic Management, Kortrijk, Belgium (2005); Integrated Vessel Traffic Management and Information Systems for port, waterway, coastal and maritime security, Port Technology International 28th edition.

Capt. Mark Johnson (1999); Proceeding of Marine Safety council – October-December 1999, Waterways Management: A New Coast Guard Business Line.

Captain B. Sitki Ustaoglu And Masao Furusho (2002); *The Importance And Contributions Of The VTS Towards The Establishment Of The Global Safety Management System For The Safety Of The Maritime Transportation.*

Captain Björn Lager (1996); *Procurement of a new VTS: Some essential factors,* Port Technology International Magazine, 1996.

Dr. Capt. Ender Asyali (2002); A Ship Based Approach to Determine the Effectiveness of VTS Systems in Reducing Vessel Accidents in Congested Waters, Dokuz Eylul University, Turkey.

Gold, E. (1990);: Legal and Liability Aspect of VTS Systems – *Who Is To Blame When Things Go Wrong*, The Nautical Institute On Pilotage and Shiphandling, 216-221.

IMO Resolution A.578 (14), (1985); Guidelines for Vessel Traffic Services, 20 November 1985.

IMO Resolution A.857 (20), (1997); Guidelines for Vessel Traffic Services, 27 November 1997.

Ingo Harre, STN ATLAS Elektronik, Bremen (2001); An Integrated Approach to Vessel Traffic Management and Information Services.

International Convention for the Safety of Life at Sea (SO-LAS), 1974; http://www.imo.org/Conventions/contents.asp?topic_id=257&doc_id=647 – Cited 1st September 2009.

J. Carson-Jackson (2005); AIS: A tool for VTS, Port Technology International 27th edition.

Lieutenant Michael Day and Thomas Wakeman III (1999); Proceedings of the Marine Safety Council, October-December 1999; Meeting the Challenge of An Integrated Information System in the Port Of New York and New Jersey.

National Research Council (1996); Vessel Navigation And Traffic Services For Safe And Efficient Ports And Waterways, Committee on Maritime Advanced Information Systems, Marine Board Interim Report, National Academy Press, Washington, 1996.

National Research Council (1999); Applying Advanced Information Systems to Ports and Waterways Management, Committee on Maritime Advanced Information Systems, Marine Board, Commission on Engineering and Technical System, National Academy Press, Washington, 1999.

Ric Walker (1999); Waterways Management Research & Development.

The International Association of Marine Aids to Navigation Lighthouse Authorities (IALA) (2002); *IALA Vessel Traffic Services Manual (VTS) Manual.*

Transas Ltd (1995); Vessel traffic control as value added management, Port Technology International Magazine, 1995.

23. A Simulation Environment for Modelling and Analysis of the Distribution of Shore Observatory Stations - Preliminary Results

T. Neumann
Gdynia Maritime University, Gdynia, Poland

ABSTRACT: The paper has presented the usage of mathematical theory of evidence in evaluating of the possibility of object detection by monitoring radar stations. The level of object detection allows for effortless conversion to optimisation problem of monitored area coverage. Development of such task enables such distribution of observatory stations that maintains the detection rate higher than the assumed value. An appropriate rate level is achieved by covering the analysed set of points with sufficient number of radar stations. Combining evidence allows for calculating corresponding parameters for each set of observing equipment.

1 INTRODUCTION

A highly significant issue during the planning and construction of the shore observatory stations network is undoubtedly their proper location, as the success of the entire investment depends on the right positioning. The criterion for project evaluation may be adopted on the basis of the extent of coverage of the monitored area, the number of observatory stations used for this purpose and, hence, the degree of maritime transport safety. Restrictions imposed on such dilemma require a thorough analysis of the issue long before the realisation of the project. Problem analysis ought to focus on the placement of shore stations primarily. During the problem analysis various possible locations and observatory station types should be regarded. The results of completed analysis should answer the question which of the possible locations are the best for creating the sufficient network of the shore observatory stations.

Above all, the choice of location ought to fulfil marine shipping safety requirements, which are to a great extent related to the warranty for the monitored area coverage, as well as various technical and economic aspects. Provided the system constructed is a mere expansion of an already existing maritime traffic monitoring system, it ought to take into account existing observatory infrastructure.

2 FACILITY LOCATION AND LOCATION SCIENCE

Facility location problems investigate where to physically locate a set of facilities (resources, stations, etc.) so as to minimize the cost of satisfying some set of demands (customers) subject to some set of constraints. Location decisions are integral to a particular system's ability to satisfy its demands in an efficient manner. In addition, because these decisions can have lasting impacts, facility location decisions will also affect the system's flexibility to meet these demands as they evolve over time.

Facility location models are used in a wide variety of applications. These include, but are not limited to, locating warehouses within a supply chain to minimize the average time to market, locating hazardous material sites to minimize exposure to the public, locating railroad stations to minimize the variability of delivery schedules, locating automatic teller machines to best serve the bank's customers, locating a coastal search and rescue station to minimize the maximum response time to maritime accidents, and locating a observatory stations to cover monitored area. These six problems fall under the realm of facility location research, yet they all have different objective functions. Indeed, facility location models can differ in their objective function, the distance metric applied, the number and size of the facilities to locate, and several other decision indices. Depending on the specific application, inclusion and consideration of these various indices in the problem formulation will lead to very different location models (Hale & Moberg, 2003).

There exists two predominant objective functions in location science: minisum and minimax. These are also known as the median and centre problems, respectively. The diametrics of these objective functions also exist (maxisum and maximin), although they are somewhat less studied. Other objective functions are also studied within the location science community, especially recently. The most notable of these are the set covering and maximal covering objective functions. The former of these two objectives attempts to locate the minimum number of new facilities such that a prescribed distance constraint to existing facilities is not violated. In contrast, the latter strives to locate a given number of facilities to best meet the (weighted) demands of the existing facilities subject to a maximum distance between new and existing facility. It should be noted that for the set covering formulation, because all of the demands must be met (covered) regardless, the relative weight of the demands generated by the existing facilities are inconsequential, whereas in the maximal covering objective some existing facility demands may be left unmet (uncovered).

Location problems are generally solved on one of three basic spaces: continuous spaces (spatial), discrete spaces, and network spaces. The first of these three deals with location problems on a continuous space (in one, two, or three dimensions) where any location within the realm is a feasible location for a new facility. The second looks at problems where locations must be chosen from a pre-defined set while the third looks at location problems that are confined to the arcs and nodes of an underlying network (Hale, Moberg, 2003).

3 MATHEMATICAL THEORY OF EVIDENCE IN MARINE TRAFFIC ENGINEERING

3.1 *The classical approach*

The mathematical theory of evidence, one of various tools employed in the application, will be used for evaluating the possibility of objects detection by, e.g., radio stations monitoring specified area. Approximate reasoning, the other significant element used in the application, provides extrapolation and interpolation of the incomplete knowledge of monitoring stations characteristics. The solution to the problem of detection, namely assessing the possibility of detection by each station, allows for moving on to the optimisation problem of supervised area coverage. Such task can be solved by locating observatory stations in a manner enabling detection level indicators to be higher than the assumed value. An adequate level of indicator values is achieved by covering a set of analyzed points with the relevant number of suitably located stations. Detection levels are estimated for a specific discrete search area with regard to records consisting of observing equipment data. Submission of records allows for calculating appropriate parameters for each point of the analyzed area. Solving this problem is possible through the mathematical theory of evidence.

The characteristics of each observatory station are the starting point in the computational process. Sufficient technical parameters provided by the equipment manufacturers are incomplete values, calculated for a certain meteorological conditions at sea (e.g., calm sea) and for typical naval units. Approximate inference mechanisms as well as additional knowledge of experts are needed, the latter being a major source of output data due to the subjective expert assessment, affected by a degree of uncertainty. This subjectivism requires the use of appropriate mathematical apparatus. The mathema-tical theory of evidence in its flexibility allows the use of fuzzy and approximate figures. Calculation of masses of the particular framework of discourse hypotheses with use of membership function creates a structure of beliefs enabling extensive use of the mathematical theory of evidence.

Water areas monitoring is conducted with the use of radar (among others), characterized, as any other technical equipment, by certain limited level of functioning and of reliability in the sense of realisation of basic tasks. The possibility of detecting object is vital parameter of any device of this kind. Modern monitoring stations are able to detect floating objects in considerable distances. One may venture to say that their range is horizontal in fine hydrometeorological conditions. Obviously, the ability to detect depends on the so-called Radar Cross Section (RCS), a feature which is associated primarily with the size of a given unit. Another significant parameter is a draft of a vessel, which in turn reduces the above-water body, affecting directly the size of the reflecting surface. Detection of smaller units at rough sea may pose problems. Whether a specific type unit will be detected in particular conditions depends on the distance to the observatory station, the sea state and the observing equipment characteristics. In some areas the heavy maritime traffic calls for an appropriate location of observatory stations ensuring sufficiently high level of unit detection.

The Dempster–Shafer theory (DST) is a mathematical theory of evidence. In Dempster-Shafer (DS) theory, there is a fixed set of n mutually exclusive and exhaustive elements, called the frame of discernment, which is symbolized by:

$$\Omega = \{A_1, A_2, ..., A_n\} \qquad (1)$$

The representation scheme Ω describes the working space for the desired application since it consists of all propositions for which the information sources can provide evidence. Information sources can distribute mass values on subsets of the frame of discernment, $A_i \in 2^\Omega$. An information source assign mass values only to those hypotheses, for which it has direct evidence.

$$0 \le m(A_i) \le 1 \qquad (2)$$

Basic Probability Assignment (BPA) has to fulfil the conditions as follows:

$$m(\phi) = 0 \qquad (3)$$

$$\sum_{A_i \in 2^\Omega} m(A_i) = 1 \qquad (4)$$

If an information source cannot distinguish between two propositions, A_i and A_j, it assigns a mass value to their union $(A_i \cup A_j)$. Mass distribution from different information sources, are combined with Dempster's rule of combination (5). The result is a new distribution, which incorporates the joint information provided by the sources.

$$m(A_k) = \frac{\sum_{A_1 \cap A_2 \cap ... \cap A_d = A_k} \left(\prod_{1 \le j \le d} m_j(A_j) \right)}{1 - K} \qquad (5)$$

$$K = \sum_{A_1 \cap A_2 \cap ... \cap A_d = \phi} \left(\prod_{1 \le j \le d} m_j(A_j) \right) \qquad (6)$$

A factor K is often interpreted as a measure of conflict between the different sources (6) and is introduced as a normalization factor (5). The larger K is when the more the sources are conflicting and the less sense has their combination.

If factor K = 0, this shows a consensus of opinions, and if 0 < K < 1, it shows partial compatibility.

Finally, the Dempster's rule of combination does not exist when K = 1. In this case, the sources are totally contradictory, and it is no longer possible to combine them. In the cases of sources highly conflicting, the normalisation used in the Dempster's combination rule can be mistaking.

From a mass distribution, numerical values can be calculated that characterize the uncertainty and the support of certain hypotheses. Belief (7) measures the minimum or necessary support whereas plausibility (8) reflects the maximum or potential support for that hypothesis.

$$Bel(A_i) = \sum_{A_j \subseteq A_i} m(A_j) \qquad (7)$$

$$Pl(A_i) = \sum_{A_j \cap A_i \ne \phi} m(A_j) \qquad (8)$$

3.2 *Mathematical theory of evidence and fuzzy values*

When the assessment of the situation undergoes solely a subjective expert rating, the results are only to be obtained in form of linguistic variables. Theories presented show (Zadeh, 1975) possibility of transforming such values into figures with use of the fuzzy sets theory, a concept created by L.A. Zadeh in the sixties of the 20th century and developed ever since (mainly by its author), which increasingly intercedes in various economic issues. According to Zadeh, the aforementioned theory has not been sufficiently employed for the purpose of detection analysis of marine units. A more extensive use of possibilities offered by the fuzzy sets theory appears as a necessity for rational construction of new maritime traffic monitoring systems.

The mathematical theory of evidence deals with function combining information contained in two sets of assignments, subjective expert ratings. This process may be interpreted as a knowledge update. Combining sets results in forming of new subsets of possible hypotheses with new values characterising probability of specific options occurrence. The aforementioned process may continue as long as provided with new propositions. This function is known as Dempster's rule of combination.

A fuzzy nature can be attributed to events which may be interpreted in fuzzy manner, for instance, inaccurate evaluations of precisely specified distances to any point. Subjective evaluations in categories: near, far, very far may be expressed with fuzzy sets defined by expert opinions. Such understanding of fuzzy events is natural and common. Introduction of events described by fuzzy sets moderates the manner in which the results of processing are used, expands the versatility of such approach, as well as changes the mode of perceiving the overall combining procedure. Deduction of specific events involved in the process of combining pales into insignificance, as obtaining information on related hypotheses is of greater interest. Combining evidence of fuzzy values brings new quality into knowledge acquisition due to the usage of combination results as a data base capable of answering various questions. After combing many fuzzy distances, the results allow to set the support level for the veracity of statement claiming a distance between a vessel and a barrier is very close, safe or yet another. Other possibilities of the mathematical theory of evidence in problems of navigation can be found in Filipowicz, 2010.

As to the problem of monitored area coverage, phrases used for assessing the distance of units will be linguistically interpreted. The determination of distance at which an object is located is possible with use of linguistic values: very close, close, far, very far. Particular linguistic expressions and corresponding exemplary range of values are presented in the Figure 1. More on this subject can be found in Neumann, 2009.

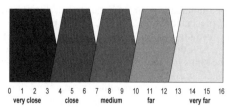

0 1 2 3 4 5 6 7 8 9 10 11 12 13 14 15 16
very close   close   medium   far   very far

Figure 1. Diagram of linguistic relations

3.3 *The optimisation problem of observatory stations distribution*

Optimisation is determining the finest solution, therefore finding the extremum of given function in terms of specified criterion (e.g., cost, time, distance, efficiency).

The optimisation of observatory stations distribution is achieved by such location of stations that makes the tracking surface of the monitored area nearest to the overall surface of study area. Obtaining equality in two aforementioned surfaces marks an ideal state of no occurrence of shaded area. The usage of a given sort of observatory stations affects the type and size of shaded areas resulting from environmental impact. In case of conventional radar stations tracking surface depends on:

– blind spots caused by port infrastructure, topography of the area,
– range and bearing discrimination.

Arrangement of shore stations seeks to maximise the tracking surface while minimising the number of newly built radar stations. Object-caused bind spots may be eliminated by installing higher number of observatory stations in various locations in a manner enabling their range to cover the entire tracking surface. This goal may be achieved by using a number of additional observatory stations. However, one of the regarded optimisation criteria are economic conditions, additionally: the characteristics of the terrain where stations are to be built, technical possibilities of connecting stations to the network. The presented approach has been implemented in the computer application allowing for analysing maritime areas, proposing a distribution of observatory stations, as well as for comparative assessment of existing area monitoring schemes.

4 THE FEATURES OF APPLICATION

The application was created for MS Windows operating system (Neumann, 2010). This software provides an uncomplicated system for management of stored shoreline patterns along with the suggested locations of shore stations. This very construction scheme for the aforementioned application was chosen due to its ability to design, analyze and assess practically all substantial parameters to be evaluated.

Its functions include: creating new projects, browsing the list of existing projects, browsing the information on given project and editing information on given project. Transparent, clear and flexible application architecture makes expanding the system and implementing changes easier. This allows to avoid many mistakes that could have a significant impact on the quality of performed calculation.

The application was designed according to the design pattern Model-View-Controller, with the main purpose of separating the part of application responsible for realisation of data processing from the representative part displaying data. Such partition has numerous advantages, e.g., greater flexibility of the application, more readable code and increased degree of code reuse.

In such constructed architecture three main layers may be distinguished:

– the presentation layer – responsible for output data formatting and displaying the final result,
– the business logic layer - mechanisms employing the basic logic of the application with implemented methods of mathematical theory of evidence. Contain the entire application engine.
– the data layer – the database and the data stored in the system.

The main purpose of this application is such distribution of shore observatory stations that provides coverage of the entire desired area. Therefore, principal objectives of the observatory stations location test may be specified as follows:

– to ensure the greatest possible coverage of the monitored area, using the smallest possible number of observatory stations.
– to ensure the highest possible probability of floating unit detection in the medium weather conditions.

The greatest possible coverage is one of the input parameters of the application. While using the visual interface of the application, the user determines the maritime area of his or her interest, whose monitoring is to be ensured by the shore stations. The user determines also the set of localisation where shore stations may be constructed. A unit detection characteristics in inclement, medium and fair weather conditions may be entered for each station. The entered characteristic is converted to the corresponding values of detection possibility for the specified point. Calculated values can be changed at any moment of application's work, hence the determination of, e.g., permanent dead zones for observatory stations becomes possible.

One of the application elements is the module, which allows for defining detailed characteristics of an observatory station. By defining a station, the user determines its maximum range of floating unit detection. Characteristics for each station ought to be created with regard to all features affecting the quality of observation. The method for defining parameters is presented in the Figure 2. From the viewpoint of the brick defining individual intervals and the level of medium size floating unit detection in medium weather conditions are highly significant elements. The employment of linguistic operator will improve work with the application and emulate behaviour similar to human reasoning. Entering any directional characteristics for each station gives the opportunity to accurately reproduce the actual situation, increase range of the station in the direction of fine terrain parameters, while limiting the range where natural barriers preventing good observation occur.

The main algorithm for this application analyses each of the interesting points with use of the mathematical theory of evidence, searching for such distribution of observatory stations, that would ensure the specified coverage of the given area. Of all locations available the smallest number is chosen for observatory stations.

An entirely separate module is the comparative analysis of various observatory stations arrangements, allowing for choosing the location for an observatory station, providing its characteristics and, subsequently, calculating the detection level in specific points in the chosen area. The calculation results for each observatory stations distribution system are collected in detailed report specifying additionally the best of given shore stations distribution systems, as well.

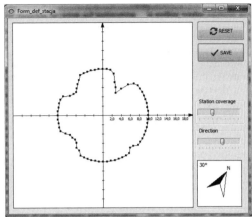

Figure 2. Module for defining detailed characteristics of an observatory station.

In every simulation project adapting the general model to the specific problem situation plays essential role. Adapting particular model of simulation project realization does not equal the lack of possibility to introduce changes, contrary – it ought to be continuously modified to consider conditionings of specific projects.

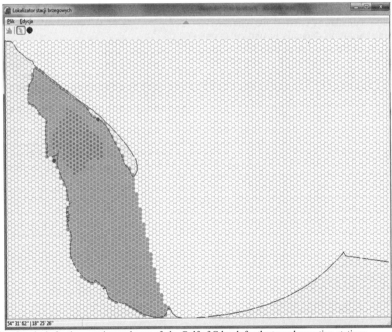

Figure 3. Results of analysis for the experimental area of the Gulf of Gdansk for the two observation stations

At the current stage of work on creating simulation environment, the completion of calculations depends on fulfilling weather condition for each of analysed points. The method suggested enabling assessment of observatory station emplacement searches in the solution set for the solution assuring unit detection at the level higher than the entered value. The number of solutions to the problem can be bigger, yet it is always a finite number. All of solutions obtained in this way may be compared. The solution in which the sum of all calculated values of specific points detection coefficients has the highest value may be indicated as the best one. As employing this method as the best point at the solution of extreme values, an alternative way for choosing the solution (based on assigning weights to possible results intervals and calculating the sum of coefficients with regard to aforementioned weights) was implemented.

The results analysis may suggest changes in shore observatory stations distribution. Other location of observatory stations can cause enlargement of the observation area in monitored maritime areas and a rise in the number of current situation data. More effective distribution of stations contributes to the navigation safety improvement and brings measureable benefits to the environment, as every failure to detect a floating unit may result in shipwreck and contamination of the environment.

The visualisation of the results is presented in the Figure 3. Degrees of colour saturation stand for values obtained in specific points of analysed problem.

The method for calculating coefficients characterizing the obtained solution depends in particular on the representation of the assessments of events occurring in the defined structures. The assessments may have form of exact or approximate values; it may be based on the interval values or fuzzy values. The type of used assessment requires employment of adequate mathematical apparatus when combining evidence.

5 SUMMARY

The paper has presented the usage of mathematical theory of evidence in evaluating of the possibility of object detection by monitoring radar stations. The level of object detection allows for effortless conversion to optimisation problem of monitored area coverage. Development of such task enables such distribution of observatory stations that maintains the detection rate higher than the assumed value. An appropriate rate level is achieved by covering the analysed set of points with sufficient number of radar stations. Combining evidence allows for calculating corresponding parameters for each set of observing equipment.

The usage of software engineering methods in simulation study on distribution of shore observatory stations is not limited solely to the possibility of realising a project of such type on the basis of model derived from this field. The growing importance of IT factor entails verification of models in a documented and structured manner – therefore techniques developed for software testing are also applicable. The main purpose of the application is to determine locations of shore observation points in order to ensure the control, management and maritime traffic control in areas of limited surface.

REFERENCES

Filipowicz, W. 2010. Fuzzy Reasoning Algorithms for Position Fixing, Pomiary Automatyka Kontrola 2010(12), s. 1491-1494

Hale, T.S. & Moberg, C.R. 2003. Location Science Research: A Review. Annals of Operations Research 123, 21–35, Kluwer Academic Publishers, The Netherlands.

Neumann, T. 2009. Problemy Wyznaczania Lokalizacji Brzegowych Stacji Obserwacyjnych, TransComp, Zakopane (in Polish)

Neumann, T. 2010. Komputerowe narzędzie wspomagające analizę lokalizacji stacji obserwacyjnych rejonów morskich. Logistyka, Nr 6, CD-ROM (in Polish)

Zadeh, L.A. 1975. The concept of a linguistic variable and its application to approximate reasoning, Inf. Sci.

24. The Relation with Width of Fairway and Marine Traffic Flow

Yu Chang Seong, Jung Sik Jeong & Gyei-Kark Park
Faculty of division of maritime transport system, Mokpo Maritime University

ABSTRACT: Nowadays, many of changes have arisen in maritime traffic due to the enlargement of ship's size and improvement of ship's speed. It is common that the risks of handling a ship in narrow water area is increasing according to increase traffic volume. In order to correspond to these changes and risks, it should be necessary to make sure the relation between proper width of route in fairway and allowable traffic volume. The conventional method of designing a fairway takes into consideration of movement by one ship of maximum size. But the congestion by traffic volume is not taken in the concept. In a while, a case of road design is generally considered about a traffic volume. From such a view point, this research proposes the method of determining fairway-width in consideration of traffic flow. To evaluate traffic congestion in a route, the Environmental Stress model is adopted as the index of standard, using traffic simulation with avoiding a collision for reproducing traffic flow.

1 INTRODUCTION

For two decades, many of changes have arisen in sizes and shapes of a ship in maritime traffic. Since many ships are navigating inside of narrow fairway in a harbor area specially, the navigational difficulty by traffic congestion is also increasing. In order to correspond to such a change and to secure the safety of navigation, it is required to presume the relation between a width of route in fairway and allowable traffic volume.

Generally, a fairway can be explained to be a maritime passage where ships cruise and maneuver. It can classify into ocean route and coastal route as like narrow water service as like Incheon entrance located westly in Korea. The base elements which constitute a route in fairway are a width, a length, and a depth. As usual, because the length of a route changes according to each geographical condition, there is no standard uniform in a fairway or route. However, a width and a depth are closely concerned with the safe navigation of a ship.

The existing method of route design which calculates a width is common concept that the movement of a maximum size ship in the route has taken into consideration for designing a route, but taken into consideration for another ships passing around. On other hand, in case of road design in land transport, numbers and widths of a lane are decided in consideration for designed traffic volume.

This research aims at approaching the design in consideration of traffic flow on a route width from such a view point, concurrently on the assumption that the design of route which took the movement of the one maximum model of a ship into consideration. In addition as the first approach, this research examines for one-way route in comparatively large water area which is not influenced by an island and a shoal, which has sufficient margin depth of water too.

2 NECESSITY FOR APPROPRIATE WIDTH TO CORRESPOND TRAFFIC FLOW

As a conventional method determining route width, "Approach Channels A Guide for Design" (It calls to PIANC-Rule below) is internationally proposed by PIANC and IAPH in 1997.

PIANC Rule considers a wind, a current and navigational aids such as a beacon for calculated elements in design of the width. The required width of route is needed to the multiple numbers of maximum size ship's beam (B) and a curvature radius is expressed with the multiple numbers of lengths (L) of the ship. Contrarily when it is calculated for movement of the one maximum model of ship which cruises a route, it has considered neither passing nor the existence of another ships which carries out simultaneously is taken into consideration.

In actual, various ships exist in the route in which route's state is very crowded with traffic rushes depending on time zone.

The method such as PIANC rule designs a route only depending on the basis of maneuvering maximum size model. Therefore it cannot be guaranteed of the navigational safety in such a situation.

This research will examine required width of a route from a viewpoint of navigational safety. For this examination, it is necessary to clarity two points. First is that one ship of the maximum model can guarantee safety up to how much traffic congestions become. Second point is how much width of route is needed to the capable degree of traffic volume.

3 SIMULATION OF MARINE TRAFFIC

3.1 Algorithm of Collision Avoiding

In order to calculate the width of a route in consideration of a congestion of traffic flow, it is required to reproduce a traffic congestion state under the conditions of certain given width of route, to evaluate the navigation difficulty. Here, the simulation technique based on desirability of collision avoiding was used in this research.

As an evaluation index showing the navigation difficulty under traffic congestion, the environmental stress value by an ES(Environmental Stress) model was adopted. The method of a judgment and a rule of collision avoiding in the simulation model used here is explained below.

3.1.1 Judgment of collision avoiding>

A start domain of collision avoiding is a standard area which sets up the start of collision avoiding according to approach a collision distance by another (target) ship, just when there is a possibility of colliding with ownship.

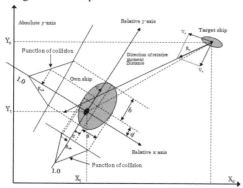

Fig.1 Concept of Collision Avoidance

In the model used in this paper, it also sets to the relative speed of an ownship and target ships according to average lengths among the ships, as shown in Fig.1.

3.1.2 Process of collision avoiding

As shown in a formula (1), the efficiency of collision avoiding and the safety of ownship are calculated by all navigational means including change a course and a speed. The optimal area finally is chosen by an evaluation.

$$u(X_{i,j}) = Pb(X_{i,j}) - a \times \underset{K=1,m}{Max}\{R_k(X_{i,j})\} \qquad (1)$$

$u(X_{i,j})$: Evaluation Function to choose optimal navigation

$Pb(X_{i,j})$: Probability changing all courses and speeds to avoid

$R_k(X_{i,j})$: Risk colliding with ships or obstacles

$X_{i,j}$: Area to avoid collision with other ships or obstacles

$i = 1 \sim p$: Preference group to change a course

$j = 1 \sim q$: Preference group to change a speed

a: Weight factor, m: Ship's numbers of encountering

Probability changing a course to avoid:
$Pb(X_{i,0}) = \exp(-a_c \cdot \Delta C_0)$

Probability changing a speed to avoid:
$Pb(X_{0,j}) = \exp(-a_v \cdot \Delta V)$

Probability changing a courses and a speed to avoid:
$Pb(X_{i,j}) = Pb(X_{i,0}) \cdot Pb(X_{0,j})$

ΔC_0 : An angle of course change (deg)

ΔV : A ratio of speed change (%)

$a_c = 0.0190$(Left change of course), 0.0260(Light change of course), $a_v = 0.0456$

3.2 Condition of simulation

A simulation area is considered to one-way passage, which the length is 10000m. The evaluation section is 4000m off the center of this area. This water area is not influenced by an island, a shoal, etc, which has also sufficient depth of water.

About a condition of a simulation, three components were set up parametrically; a width of route, composition of ship's appearance and traffic volume as shown in Fig.2 Process simulation.

The width of a route was considered as three patterns (300m, 500m, and 700m) by referring to main ports of Korea. The compositions of ship which form a traffic flow set up for small size, medium size and large size patterned to be a percentage of 7:2:1 and 4:5:1. Traffic volumes could be 10, 20 and 30 ships per an hour.

The size of a ship was classified into the small ship (48.26m±20m), middle ship (104.08m±20m) and the large ship (240.00m±50m) based on the actual data of an entrance to Mokpo in Korea. The ship's speed was used as the small ship (9.7kts±2.2kts) and middle ship (14.5kts±3.2kts) and the large ship (15.4kts±3.0kts).

Simulation time is recommended that the time be longer in order to obtain a reliable evaluation result. However, a simulation by increasing time recklessly will be wasted. So this simulation could be up to 100 hours.

In addition, the interval time generating ships (making traffic volume) asks for the average time interval according to the number of target ships. The ships were generated using exponential distribution with such an average value.

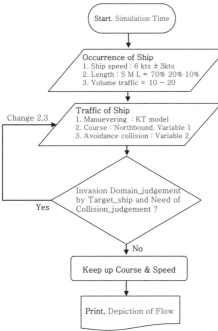

Fig.2 Process of Simulation.

4 ANALYSIS ON SIMULATION'S RESULTS

4.1 *Index of evaluation*

In this research, the environmental stress value by an ES model was adopted as an evaluation index showing the navigation difficulty under traffic congestion. A description is simply added to below about the ES model.

4.1.1 *Environmental Stress model, ES model*

The elements of the environmental conditions that can be taken into account in the model are as follows;

1 Topographical conditions such as land, shoals, shore protection, breakwaters, buoys, fishing nets, moored ships and other fixed or floating obstacles.
2 Traffic conditions such as the density of other ships and traffic flow.
3 External disturbances such as winds and currents.

The proposed model, which expresses in quantitative terms the degree of stress imposed by topographical and traffic environments on the mariner, is called the Environmental Stress Model (ES-model). The ES-model is composed of the following three parts:

1 Evaluation of ship-handling difficulty arising from restrictions to the water area available for maneuvering. A quantitative index expressing the degree of stress forced on the mariner by topographical restrictions (ES_L value) is calculated on the basis of the time to collision (TTC) with any obstacles.
2 Evaluation of ship-handling difficulty arising from restrictions on the freedom to make collision-avoidance maneuvers. A quantitative index expressing the degree of stress forced on the mariner by traffic congestion (ES_S value) is calculated on the basis of the time to collision (TTC) with other ships.
3 Aggregate evaluation of ship-handling difficulty forced by both the topographical and traffic environments, in which the stress value(ES_A value) is derived by superimposing the value ES_L and the value ES_S.

The model is a practical method for evaluating the ship-handling difficulty of navigation in topographically restricted and congested waterways, and in ports and harbors. The strength of the model lays in its ability to evaluate simultaneously or individually the difficulties of ship-handling arising from topographical restrictions and encounters with other ships and because it includes acceptance criteria based on a mariner's perception of safety.

4.2 *The standard of allowance judgment*

The marine traffic simulation with collision avoiding was carried out. The environmental stress model was applied to all the ships which form a traffic flow in a simulation. The difficulty which a ship operator felt in combinative conditions was calculated by this model.

In evaluation, it adopted as an index of a judgment of the appearance ratio [P (ESA>=890)] of the environmental stress value 890 equivalent to the state where 80 percent of ship operator groups are nonpermissible.

In addition to the judgment of the limit of permissible traffic volume, P (ESA>=890) of the ship for evaluation was made into less than 5% of the permissible judging standard, from the meaning in consideration of existence of the uncertain factor which cannot be specified.

4.3 Result of simulation

4.3.1 Calculation of allowable traffic on route width

In evaluation of ship traffic, it could not consider that a small ship and a large ship were the same traffic of 1 vessel, but traffic was normalized using L conversion traffic by using a 70m [in full length] ship as a standard ship.

Fig.3 compares route width with the appearance ratio of load nonpermissible for every combination of an ownship model. In simulation, it has a plot which case route width is from 300m to 700m.

In these figures, ship's percentages are <4:5:1> and <7:2:1> is united and plotted. Since these ships were expected that they could normalize using L conversion traffic, there were some variations. But the graphs performed linear regressions for the group of six points respectively.

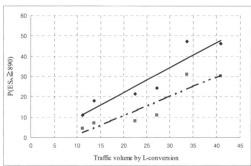

Fig. 3 Rate of appearance of P (ESA>=890) by width of route in 500m (◆: Large ownship, ■: Middle ownship)

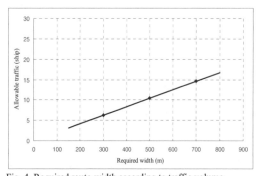

Fig. 4. Required route-width according to traffic volume (In case ownship is middle-sized)

Although an ownship is extrapolated as a part, the amount of allowable traffic to each width of route and each case of ownship's size can be obtained if an approximation straight line reads L conversion traffic in case the value of a vertical axis [P (ESA>=890)] used as 5% based on these figures.

Thus, the amount of allowable traffic to each route width can be presumed from Fig.4 when an ownship are each a large size and a middle size ship.

4.3.2 Presumption of allowable traffic on route width

Fig.4 shows plotted graphs presumed result of the amount of allowable traffic to each route width when taking from the case in a middle size of ownship.

From this figure, the required route width to the allowable limit of the traffic can be read to the given route width conditions or the traffic conditions.

If it becomes route to 500m for a large ship, 10 ships per an hour will become as for the amount of allowable traffic of which L conversion was done. If 700m, 15 ships per an hour will become.

When it comes from a middle size ship, it turns out that permissible traffic volume increases respectively by every 5 ships per an hour.

4.4 Relation to existing rule of route design

The existing route design method is a common method of designing in consideration of movement of one maximum ship in a route.

However on an actual route, the states that ships navigate to compete with other ships would be a normal state. In this research, the design of the route width in consideration of traffic was approached, escaped from the conventional route width design method which is width in consideration of movement of the one maximum model of a ship.

Fig.5 is a mimetic diagram shows the mutual relation of the design method in consideration of the existing route width design method and the congestion of traffic based on Fig.4.

If the route width designed by the existing method cannot allows a certain traffic to exceed, it can explain the need of increment of fairway which the concept of shifting to the view based on the route width in consideration of traffic volume designed by this research.

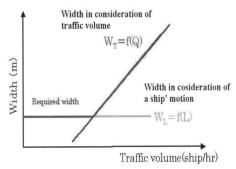

Fig. 5, Idea of required route-width over traffic volume

5 CONCLUSION

As many changes in maritime conditions have arisen in narrow waterway including the enlargement of ships, so navigational difficulty by traffic confusion is increasing. In order to secure the safety of navigation to such a change, it is needed the concept based on ship's traffic, which concept considers ship's traffic in congested situation and the maneuvering movement of maximum size ship in the traffic.

From such a viewpoint, the route design method in consideration of traffic congestion was approached by presuming the relation between route width and permissible traffic amount.

This paper examined for one-way route in comparatively large water area which is not influenced by ships and any obstacles as the first approach, which has sufficient margin depth of water. Asked for appropriate width of simulated route in consideration of congested traffic, the traffic congestion state was reproduced by marine traffic simulation. The navigation difficulty under traffic confusion was evaluated with the application of Environmental Stress model.

As a result, it could clarity that how much of width in a route be able to guarantee a state of traffic congestion to navigational safety in the simulation.

REFERENCES

R.G. Garthune et al. (1948) : The Performance of Model Ships in Restricted Channels in relation to the Design of a ship Canal, DTMB report 601.

J. Siebol and J.H.G. Wright (1974) : Application of Ship Behavior Research in Port Approach, Proceedings of Ship Behavior in Confined Water, Symposium held in London.

PIANC PTC II (1997) : Approach Channels A Guide for Design. Report of Working Group II −30. Supplement to Bulletin No.95.

Kinzo Inoue et al. (1998) : Modeling of Mariner's Perception of Safety, The journal of navigation, No.98, pp. 235-245. (in Japanese)

Kinzo Inoue (2001) : Evaluation Method of Ship-handling Difficulties for Navigation in Restricted and Congested Waterways, The Journal of Royal Institute of Navigation, Vol.53, No.1, pp.167-180.

25. Integrated Vessel Traffic Control System

M. Kwiatkowski, J. Popik & W. Buszka
Telecommunication Research Institute Ltd., Gdańsk, Poland

R. Wawruch
Gdynia Maritime University, Gdynia, Poland

ABSTRACT: Paper describes Integrated Vessel Traffic Control System realizing fusion of data received from the shore based station of the Automatic Identification System (AIS) and pulse and Frequency Modulated Continuous Wave (FMCW) radars and presenting information on Electronic Navigational Chart. Additionally on the Observation Post is installed Multi Camera System consisting of daylight and thermal cameras showing automatically object tracked by radar and selected manually by operator.

1 INTRODUCTION

Paper describes Integrated Vessel Traffic Control System realizing fusion of data received from shore based station of the Automatic Identification System (AIS) and pulse and Frequency Modulated Continuous Wave (FMCW) radars and presenting information on Electronic Navigational Chart issued by the Polish National Hydrographical Service – Hydrographical Office of the Polish Navy. Additionally on the Observation Post is installed Multi Camera System consisting of daylight and thermal cameras showing automatically object tracked by radar and selected manually by operator.

System was designed and built in the scope of research work financed by the Polish Ministry of Science and Higher Education as developmental project No OR00002606 from the means for science in 2008-2010 years.

2 MODEL OF INTEGRATED VESSEL TRAFFIC CONTROL SYSTEM

2.1 *System configuration and features*

Model of Integrated Vessel Traffic Control System has a modular architecture. This system has been subdivided into autonomous modules. Each of them can be individually maintained, upgraded and/or replaced. Presented system has been designed to make possible development of his elements and functions in the future. The structure of the system has been shown in Figure 1.

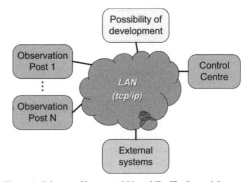

Figure 1. Scheme of Integrated Vessel Traffic Control System.

The Integrated Vessel Traffic Control System is an open-ended system, which is possible to integrate with external systems. The system consists of one Observation Post and one Control Centre. If required, there are abilities to connect to Control Centre several Observation Posts. All elements of the system are linked through the LAN (Local Area Network).

Figure 2 shows functional diagram of Observation Post. It consists of the following devices:
- Pulse radar in X band;
- Frequency Modulated Continues Wave (FMCW) radar in X band;
- Multi cameras system which comprise of daylight and thermal cameras; and
- Class A station of the Automatic Identification System (AIS).

The pulse radar as well as the FMCW radar includes plot extractor and tracker device.

The pulse NSC 25/34 Raytheon radar has been applied in the model of integrated system for the performance tests. This radar is characterized by the maximum range equal 24 Nm. However the CRM-203 FMCW radar, also used in the Observation Post, has been produced by Telecommunication Research Institute Ltd. This radar is able to detect objects at maximum distance equal 48 Nm. It is necessary to remember that range of radar destined to detection of sea surface objects, also depends on a height of antenna location and propagation conditions of electromagnetic wave.

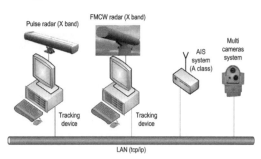

Figure 2. Functional diagram of the Observation Post.

CRM-203 is fully solid state radar. It makes use of modern technologies such as generation of emitting signals on the base of frequency Direct Digital Synthesis (DDS) and Fast Fourier Transform (FFT) processing implemented in signal processors TigerSHARC and Field Programmable Gate Array (FPGA) circuits VIRTEX. Described radar can detect sea surface targets and determine movement parameters of objects. It is equipped with modules realizing ARPA's functions and can cooperate like pulse ship's radars with external sources of information (GPS receiver, ship speed measuring device, gyrocompass, satellite compass, scanner of charts, etc.). CRM-203 is able to automatically tracking of detected targets and to pass information about tracked objects to command and control systems.

More detailed information about utilised FMCW radar was presented during the previous TransNav conference (Plata 2009).

Multi cameras system Sargas KDT-360 has been installed on Observation Post of integrated system. KDT-360 has been manufactured by the Polish company Etronika. Maximum range of the multi cameras system is in the order of several kilometers. If necessary, it is possible to apply in the integrated system cameras, which have better range features.

Moreover for performance tests purposes in Integrated Vessel Traffic Control System has been used as AIS coastal station AIS class A device produced by SAAB.

The main task performed by the Observation Post is supply information about sea surface situation in detection range of the radars and operation range of the AIS system. Moreover, remote controlled multi cameras system can make identification of objects detected by remain sensor or even external systems. All devices are able to work in the unattended mode.

The functional diagram of the Control Centre has been presented in Figure 3. The centre consists of equipment as follows:
- Server;
- Operator workstation; and
- Printer.

Server as well as operator workstation operate under control of Red Hat Enterprise Linux 5 operating system.

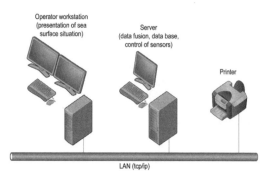

Figure 3. Functional diagram of the Control Centre.

The Dell Power Edge T610 server has been used in the presented system. This computer has two four cores and 64 bits processors Xeon X5570, 32 GB of RAM as well as five 300 GB hard disks. Hard disks are operating under matrix system RAID 6.

The Dell Precision T5500 computer has been applied as operator workstation. This machine has 64 bits processor Intel Xeon W5580, 12 GB of RAM, two 300 GB capacity hard disks and graphical card NVIDIA Quadro FX3800. Hard disks in the workstation are operating under matrix system RAID 1.

The Control Centre performs tasks as follows:
1 Generalization of information about current sea surface situation. The process relies on verification and data association collected from sensors, which operate on Observation Post.
2 Monitoring, collecting and updating of sea surface information within the scope of:
   - Tracking of all sea surface targets in range of Control Centre operation;
   - Identification and classification of detected object;
   - Distinguishing between stationary and movable objects;
   - Assertion of vessel entrance to areas temporarily or permanently prohibited and other areas defined by the operator;
   - Assertion of vessel descent from in forcing maritime routes; and
   - Cooperation with others services.

3 Archiving and play back of recorded sea surface situation.

Information received from radars and AIS are put to the database. Next, these data are subjected to fusion. Results of fusion are also written down to the base. Communication and control of the sensors, the database and the data fusion algorithms have been implemented in the server.

Data from the base and the sensors i.e. radars, AIS and even multi cameras system can be transferred automatically (in suitable range) to external systems. Information from these external systems can be received automatically, as well. Received data, such as targets tracks from radars and AIS are written down to the base and are subject to fusion with information obtained from local sensors of Integrated Vessel Traffic Control System.

Information from all local sensors and external systems are presented on two computer displays. Actual positions and movement vectors of targets, detected by radars and received from AIS devices installed on vessels, are presented in the graphical form on Electronic Chart Display and Information System (ECDIS) (Weintrit 2009). The picture from multi cameras system is presented in the separate window (daylight or thermal camera). Archived comprehensive sea surface situation can be playing back in any moment.

Presented configuration of the model of integrated system and its functions can be changed. Therefore, modifications and development of manufactured vessel traffic control system are possible in dependence on needs and requirements of a customer.

2.2 Data fusion

Multi-sensor data fusion is one of the most effective ways to solve problems of different groups, which have common characteristic features. It uses data from multiple sources to achieve a result which would not be possible to obtain from a single sensor. Data received from different sources can be associated by make use of specific procedures of signal processing, recognition, artificial intelligence and information theory. Many methods of data fusion have been developed and their common feature is an inclusion of multiple layers of data processing in the integration process.

Fusion process used in presented system can be divided into several main stages (Figure 4):
1 Unification of state vectors units of targets and bringing them to single timeline.
2 Association.
3 Determination of an updated state vectors.

The need to harmonize the time, results from the asynchronous operation of sensors. They operate with different frequencies and can be turned on at different times. For example the position of radar an-

tennas working in the system can be different at any given time. Uniform period of time equal 1 second has been adopted. It allows easy synchronization of data received from sensors with different frequencies (typically: 2, 3, 6, 10 seconds).

Association is performed using the modified PDAF algorithm (Probabilistic Data Association Filter) (Bar-Shalom et al. 1995, Bar-Shalom et al. 2001, Krenc 2006). In the modification we assume that there are two (or more) sources of varying quality. The associated measurement can come from two sources, one or none. Additionally, there is one validation gate, inside of which are measurements from both sources. The modification relay on adding new innovation vectors v and association mass e of these vectors:

$$v_{ij} = \frac{e_i \cdot v_i + e_j \cdot v_j}{e_i + e_j}, \quad e_{ij} = e^{-0.5 \cdot v_{ij} \cdot S^{-1}_{ij} \cdot v^T_{ij}} \quad (1)$$

where: v - the innovation vector; e - association mass of innovation vector; S - innovation matrix; and i, j - indexes $(i \neq j)$.

In this way has been assumed that data are received from both sensors and probability of such hypothesis is calculated. This allows several percent of improvement in the quality of estimation.

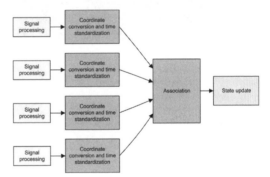

Figure 4. Diagram of data fusion process.

The state vector updates are assigned on the basis of calculated innovation vectors and association probabilities.

The algorithm allows for flexible operation, depending on currant conditions (adaptive assessment of interferences) and quantity of information sources.

2.3 Sea surface situation picture

As mentioned in paragraph 2.1, the function of presentation of sea surface situation has been implemented in the operator workstation. Picture is presented on two displays, which shows fully independent graphic information from the radars, AIS, multi cameras system and external systems. Posi-

tions and motion vectors are displayed with different accuracy, depending on the type of objects:
- Detected only by radars;
- Transmitting AIS data and not detected by radars;
- Detected by radars and transmitting AIS data, after data fusion.

Sea surface objects are presented on electronic navigation chart (Weintrit 2009). The situation can be displayed on full screen of two monitors simultaneously or one monitor or else in dedicated window. It is also possible to observe magnified area in separate window. Operator's console is presented in Figure 5. An example of costal sea surface situation picture is shown in Figure 6, whereas an example of data fusion working is presented in Figure 7.

Figure 5. Operator's console.

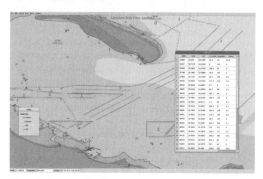

Figure 6. An example of costal sea surface situation picture.

On the left monitor there is the window displaying moving objects on the background of navigation chart. Above, on top of the screen, has been placed Menu Bar which make possible to choose the submenu for selection of actions or to open control windows. At the bottom of the screen has been placed Status Bar (see Figure 6).

The right monitor is designed generally to present windows with tables that contain descriptions of targets obtained from all sensors or external systems, additional descriptions of tracking objects and targets after data fusion as well as to present the picture

from multi cameras system. An operator can shapes the view on the screen in suitable for him manner because any windows can be moved in the frame of main window.

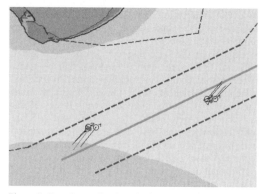

Figure 7. Two detected and tracked targets (circle 1 – FMCW radar, circle 2 – pulse radar, triangle – AIS, point – data fusion).

Software package EC2007 ECDIS Kernel developed by SevenCs GmbH has been used to build navigation charts background. The package allows the application of various types and formats of electronic digital charts supplied by different manufacturers.

The set of tools and libraries Qt developed by Nokia (formerly Trolltech) has been used to build controls elements of the picture and specialized windows, which present e.g. the list of sea surface parameters or the picture obtained from multi cameras system. Qt provides up-to-date components and mechanisms to build operator interface and tools to implement communication, processing of text documents or to use databases.

The use of both software tools, i.e. EC2007 ECDIS Kernel and Qt, allows to rapid implementation of the software which meet the requirements of IHO/IMO.

Some parts of the operator interface functions have been performed solely for research purposes, such as fusion algorithms and can not be found in the target Integrated Vessel Traffic Control System. Similarly, the units implemented in control parameters of multi cameras system are as required by this device. It has been done in order to facilitate operation tests and control of used multi cameras system. Units or scope of unit's description should be adapted to thinking ways and working procedures of an operator.

The solution of the picture displayed on monitors, which has been adopted in Vessel Traffic Control System, is flexible. Depending on the operator's needs, presentation of sea surface situation and other necessary information can be freely shaped.

3 CONCLUSIONS

Paper presents Integrated Vessel Traffic Control System designed and constructed in the scope of research work mentioned in the introduction. The system works and results of its performance tests conducted in autumn and winter 2010 are described in other paper presented on this conference.

REFERENCES

Bar-Shalom Y, Rong Li X. 1995. Multitarget-multisensor tracking: Principles and techniques. *University of Connecticut.*

Bar-Shalom Y., Rong Li X., Kirubarajan T. 2001. Estimation with Applications to Tracking and Navigation. *John Wiley & Sons Inc.*

Krenc K. 2006. Statistic methods of processed data evaluation for the purpose of information fusion in C&C systems. *MAST.*

Plata S., Wawruch R. 2009. CRM-203 Type Frequency Modulated Continuous Wave (FMCW) Radar. In Adam Weintrit (ed.) *Marine Navigation and Safety of Sea Transportation. CTC Press, Taylor & Francis Group.*

Weintrit A. 2009. The Electronic Chart Display and Information System (ECDIS). An Operational Handbook. *CRC Press, Taylor & Francis Group.*

Inland Navigation

26. Navigation Data Transmission in the RIS System

A. Lisaj
Maritime University of Szczecin, Szczecin, Poland

ABSTRACT: The paper presents processing and transmission navigation data in inland navigation. Data transmission of messages by ship electronic reporting in the RIS (River Information Services) system is analyzed. The principles of reporting data navigation in the RIS Management Centre are defined. Furthermore, the RIS communication architecture is analyzed. Finally, the author presented new standard – BICS 2.0 for the processing and navigation data transmission in inland navigation

1 INLAND NAVIGATIONAL DATA TRANSMISSION BY THE SHIP ELECTRONIC REPORTING IN THE RIVER INFORMATION SERVICES

Data transmission of messages by ship reporting in the RIS system Management Centre has following tasks [2.4]:
1 Facilitation of data structure transfer with EDI (Electronic Data Interchange) standards.
2 Exchange of information between inland navigation partners.
3 Sending dynamic information on a voyage at the same time to many participants.
4 Consistent use of the EDIFAC standard (Electronic Data Interchange for Administration)
5 Inland traffic management.
6 Transfer of complete information on locks and bridges and calamity situation.
7 Loading / unloading management and container terminal operation monitoring.
8 Border crossing control.
9 Services to passengers of inland ships.

1.1 The principles of co-operation between the RIS System Management Centre and an inland VTS stations [2.7].

General purpose of VTS center is to improve the safety and efficiency of sea and inland navigation, safety of life and the protection of the environment from possible adverse effects of vessel traffic.

A part of these objectives of a Vessel Traffic Service include promoting efficient transport and the collection of data that may be required in order to evaluate the Vessel Traffic Service.

The benefits of implementing a VTS are that it allows identification and monitoring of vessels, strategic planning of vessel movements and provision of navigational information and assistance.

It can also assist in reducing the risk of pollution and coordinating pollution response. The efficiency of a VTS will depend on the reliability and continuity of communications and on the ability to provide concise, accurate and unambiguous information.

The quality of accident prevention measures will depend on the capability of the system to detect developing dangerous situations and on the ability to give timely warning of such dangers.

For inland navigation vessels cooperating with the RIS Management Centre, there is a need to harmonize inland VTS through international application on all inland waterways in a waterway system.

2 COMMUNICATION AND REPORTING

Communication between inland VTS centre and participating vessels or between participating vessels should be limited to information essential to achieve the objectives of the VTS [1.5].

Communication should be clear, easily understood by of all participants. Standard reports and phrases should be used when necessary. Where language difficulties exist, use should be made of a common language as determined by the VTS authority.

In any VTS message directed to a vessel or vessels it should be made clear whether the message contains information, advice, warning, or an instruction.

2.1 The RIS Communication Architecture[1.3];

The RIS Architecture consists of seven components:
- **Reference Model**, in which inland shipping and River Information Services is defined and the RIS architecture s build;
- **Organization Architecture**, in which the roles responsible for the use and operation of River Information Services are defined. For the roles that actually use RIS to strengthen the execution of their task in inland shipping the intended cooperation is drawn. This cooperation is the basis for the information and functional architecture.
- **Information Architecture**, in which the information exchange coming with the cooperation as drawn in the organization architecture.
- **Function Architecture**, in which the functions are derived which have to be performed to actually make the cooperation work. At this stage these functions can be performed by the responsible roles, by those roles supported by an application or autonomous by a system.
- **Data Architecture**, in which the data is defined from which the information in the information architecture can be build.
- **Physical Architecture**, in which an first example is given of the way the functions from the functional architecture can be allocated to a system.
- **Communication Architecture**, in which the link is out between the data and information on the one hand and the standardised message on the other hand.

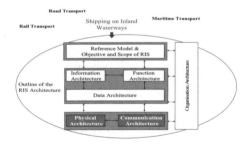

Figure 1. Structure of the data architecture in the RIS [3.6]

3 DATA TRANSMISSION IN THE RIS SYSTEM.

Technical specifications of the message structure for data processing and transmission of ship reporting in the RIS system are composed of segments shows on figure 2. The structure of a message is described in a branching diagram indicating the position and the relationship of the segments and segment groups.

For each segment the data elements are defined which are to be used in a message [3.6].

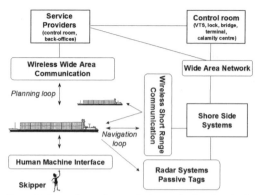

Figure 2. Processing and transmission navigation data in the RIS system [3]

3.1 New standard – BICS for the processing and navigation data transmission in inland navigation.

In conformity with the EDIFACT (Electronic Data Interchange for Administration), data processing in the RIS system makes use BICS standard [2.4].

BICS ((BICS-Binnenvaart Informatie en Communicatie System-inland shipping information and communication system)) used to transmit navigation data about transported voyages and cargos of ships.

The port authorities and all inland shipping waterway management centers need adequate information for security and safety handling of inland shipping. In the distress and emergency situations all vessels must also be able to protect people and environment.

Standard BICS transfer navigation data information faster and confidential.

BICS usually transmitted electronic declaration of ship and data cargo to the inland waterways management system.

When BICS standard is installed, all details of the ship (official of the ships number, name of the ship, dimensions and other) are defined in BICS software.

BICS contains the names of all loading and unloading points, all cargo types and the precise names of, and indications of the risks associated with, each dangerous substance that is allowed to be transported over water. It is therefore only necessary to enter the variable information for each voyage, e.g. port of departure and destination, type and amount of cargo, draught and number of persons on board.

New standard –BICS is responsible for the generation, maintenance, use and delivery processing of data. All information should be clear, understandable and logical for public authorities.

Figure 3 shows architecture of the BICS standard using on the board of the vessel.

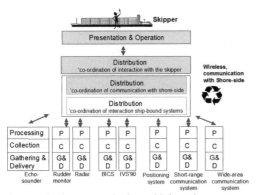

Figure 3.The data communication and information system on-board of a vessel in the BICS standard [4,8].

In line of these layers there should be a distribution layer for the co-ordination of the communication between the vessel-bound systems (e.g. echo sounder, rudder monitor, radar).

An analogous topology is drawn for the RIS-system in back-offices and control rooms cooperating with VTS-centres, locks and bridges in figure 4. For back-offices and control rooms an extra distribution layer can be of interest.

The competent authorities are allowed to distribute information based on vessels' reports to bridge and lock operators in order to optimise the traffic movements.

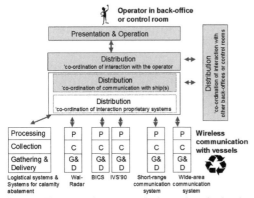

Figure 4. The processing communication and navigation information system in back-offices and the RIS control rooms management centres in the BICS standard [4.8].

4 CONCLUSION

The RIS management centres are responsibility for the storage of the all navigation data and for processing of the distribution in the BICS standard.

This standard of electronic data transmission provides rules for the interchange of electronic messages between partners in the field of inland navigation: ship owners, skippers and ports authority. For the different services and functions of River Information Services processing of the data transmission contains the most important regulations for electronic ship reporting. Standard BICS describes the messages, data items and codes to be used in the navigation data transmission.

LITERATURE

[1] Technical specifications including: inland Automatic Identification System (AIS) and International Standards for Electronic Ship Reporting in Inland Navigation. *Commission Directive (EC) no 416/2007 of 20.03.2007*

[2] Lisaj A., Electronic reporting of ships in the RIS system. [w.:] Weintrit A. (ed.): Marine Navigation and Safety of Sea Transportation. A Balkema Book. CRC Press, Taylor & Francis Group, Boca Raton – London - New York - Leiden, 2009 (TransNav 2009, Gdynia, Poland).

[3] Guidelines for planning, implementation and operation of River Information Services. *Official Journal Commission Directive (EC) no 414/2007 of 13 .03.2007.*

[4] Technical specifications for ship traffic management on harmonized River Information Services (RIS) on inland waterways. *Official Journal Commission Directive (EC) no 415/2007 of 13.03 2007.*

[5] Harre I., *AIS and VTS*. European Journal of Navigation, Vol. 1, No 1, 2003.

[6] Zieliński T.P., *Cyfrowe przetwarzanie sygnałów - od teorii do zastosowań*. WKił Warszawa 2005.

[7] Trögl J., Experiences with River Information Services in the Danube Region. Conference Smart Rivers '21, The Future of Inland Water Navigation, Vienna, Austria, 2009.

[8] http://www.bics.nl

27. River – Sea Technology in Transport of Energy Products

T. Kalina & P. Piala
University of Zilina, Slovakia

ABSTRACT: One of the key conditions to the smooth functioning of the state is its energy security. Most countries in Central and Eastern Europe are dependent on imported energy raw materials from one supplier. In connection with the growing importance of diversification of energy products, technologies and supply routes. Security of supply of strategic energy resources can be better ensured by using far less preferred alternatives. An example would be the use of inland waterways for delivery of energy products to the end consumers and enabling deliveries to places where the geographical, demographic or environmental specificities do not use traditional means.

1 INTRODUCTION

Already in 2005 the Parliamentary Committee on Economic Affairs has raised the issue of increasing Europe's energy dependence. The consequences of this dependence on imported energy resources have proven most obvious in January 2006 and 2009 when natural gas supplies from Russia via Ukraine were dramatically limited and have become an instrument of political respectively economics pressure. Even these cases confirm the fact, that energy security is one of the key conditions of smooth States functioning and globalization is essential to ensure the competitiveness of European economies. At present, only few European countries are energy self-sufficient: only Denmark, Norway, Russian Federation and United Kingdom produce more energy than they consume.

Efforts of European countries is to use as much as possible a wide range of domestic energy sources, but most of countries are reliant on imported oil and natural gas. In this context, it is a considerable problem for several countries of Central and Eastern Europe, which are absolute dependent on imported oil and gas from one supplier.

2 THE IMPORTANCE OF NATURAL GAS FOR ENERGY SECURITY OF EUROPEAN COUNTRIES

Natural gas is the world's second largest energy source. Its share in total energy consumption is now 23% per year, with its growth average of 1.6% per year. Experts estimate the state of natural gas reserves about 511 000 billion cubic meters with a lifetime of up to 200 years. However, there are three types of natural gas sources: proven natural gas reserves, probable natural gas reserves and potential natural gas reserves. There is a proven natural gas resource about 164 000 cubic meters, which its mining is currently available by nowadays economic and technical means with deliverability till 2060. 71.7% of these sources are in land and the rest 28.3% in marine shelves.

Probable reserves are reserves discovered on bearing, exhibiting a very high probability that they will be exploitable by economic and technical conditions similar to those in proven natural gas reserves. Bearings are not technically equipped yet. In addition to the category of proven natural gas reserves, with high probability, we may count with probable reserves. Just because of second resources category sectional transfer into the proven natural gas reserves, there is a still growing amount of proven natural gas reserve and its "lifetime". The probable natural gas reserves represent more than 374 000 billion cubic meters. World gas reserve (the information of the International Gas Union), taking into account consumption and deliverability of proven and probable natural gas reserves, is 12 to 152 years.

Even EU countries adapt to the general trend of replacing fossil fuels such as coal, lignite and oil well by environmentally more friendly natural gas. The Majority share of the consumed gas in Europe comes from British, Dutch, Italian, Romanian, German and Danish resources. The significant part of the total gas imports come from Russia, Norway and Algeria.

One of the aims of European policy is to diversify sources of energy and transport routes. The most discussed is just the gas. The European Union supports the construction of the Nabucco pipeline, which is considered as a strategically important energy project. Nevertheless, its development is hampered by lack of funds and weak political pressure. Nabucco should have a length of 3400 kilometers and through Turkey, Romania, Bulgaria, Hungary and Austria it should deliver annually 31 bilion cubic meters of natural gas from Central Asia to the EU. Similarly, the BTC pipeline (Baku-Tbilisi-Ceyhan), bypassing Russian territory. It should be completed in 2013, but it faces fierce competition from the planned South Stream pipeline, which develops the Russian Gazprom. The contracted amount of gas, investments, and also agreements with South stream transport countries is the project South stream considerably further than the Nabucco. Even, there were speculations that both projects could be linked together.

The Imports of Russian gas represents about 26% of the total consumption of EU countries. For Central and Eastern Europe, Russian gas is 87% of total imports. For example Slovakia, Estonia, Latvia, Finland and Lithuania depend on 100% import from Russia, Bulgaria and Czech Republic Russian gas covers 94% or 82% of their consumption. In terms of energy for the EU is the primary effort to maintain access to Algerian natural gas reserves, which could reduce dependence on Russia. Algeria's economy is heavily dependent on exports of hydrocarbons (oil and natural gas) –what make up 97% of exports, contributing 30% of GDP and finance 65% of the budget. EU imports 62.7% of Algerian exports, what is 58% of total EU natural gas imports. The weakest link in the chain of gas path from source to final consumer is a long haul. Current technology for transporting natural gas allows long distances through pipelines or tankers in liquefied form. Wide branched European network of pipelines is preferred within the continental gas transport. In the recent past, it was annexed to the undersea pipeline connecting with the sites of customers in North Africa. (Melčák, 2010)

Most gas from Algeria and Nigeria to Europe is transported in compressed form (CNG, PNG) by sea tankers into offshore terminals, followed by distribution pipelines, marine, rail and road tankers. Transhipment and storage capacity of most of these terminals is already at its limits. The solution is either construction of the new ones or substantial increase in inland traffic flows. An appropriate alternative also could be to carry liquefied natural gas (LPG).

Already today, liquefied natural gas contributed 26% of the total trade in gas. Terminals for liquefied natural gas are located in countries with large natural gas reserves. For example: Algeria, Australia, Brunei, Indonesia, Libya, Nigeria, Oman and Qatar. The whole process of natural gas liquefaction is energy-intensive. Energy needed for liquefaction of natural gas equals 1/3 energy of gas liquefaction. However, the liquefaction of gas will achieve a substantial simplification of his carriage. The big advantage is the reduction of its volume in the liquefaction process: one liter of LNG is approximately 600 liters of gas in its natural form, and low storage pressure (up to 5 bar) compared to 200 bar pressure in transport compressed gas in normal tubular pressure tanks.

In 2005, there about 50 for LPG import terminals, worldwide. The biggest receiving terminals in the world are located in Japan, which covers more than half of the global trade in LNG. In Europe, eight countries - Belgium, France, Italy, Norway, Portugal, Spain, Turkey and Great Britain has at least one terminal for processing and storage of imported LPG. France occupies the first place, which imports from Algeria 10 billion cubic meters of gas annually. Further extension of the network of European import terminals are planned or just under construction.

3 SAFETY ASPECTS OF LNG TRANSPORT

LNG transportation safety could be assessed from two views. The first is the danger of explosion and subsequent fire. The second is the environmental aspect. LNG is transported at low pressure. Because of its low temperature, the gas is transported in double-wall tanks with vacuum Perlite insulation. Perfect insulation protects contents from heat and pressure, even if the container gets into fire and lose vacuum. There are known cases where cars transporting the LNG were burnt due to a malfunction of electrical installations, but the tank remained intact. Tanks are designed according to the regulations so they withstand even external fire. There had been no accident relative to explosion or fire in the content of LNG tankers.

There are not known any maritime disasters LNG tankers, which currently operates about 200, or several dozen river LNG tankers, which operates within Europe, in contrast with oil tankers. Compared with diesel and petrol, LNG is significantly safer, but it does not mean that LNG transport is completely safe. It may occur that large LNG amounts can escape from the ship into water. In that case RPT (rapitphasetransition) effect occur. If the liquefied gas, which has a temperature of about -163 ° C will suddenly appears in a warmer ambient temperature, the liquefied natural gas will quickly change over to gas. During this transition occurs massive release of energy, which may cause an explosion.

Ignition of liquefied natural gas needs evaporation in a significant heat input and consequently it is possible to ignite its mixture with air, but only in a narrow range of concentrations from 5 to 15% at 280°C ignition, which is considerably higher value as in the case of gasoline or diesel. Prevention of such cases is associated not only with designing ships for transporting LPG, but also employing skilled crews, trained specifically for such shipments.

Neither from the environmental considerations, LNG transport does not represent increased risk. When the tanker accidents, there is not direct water damage by gas, because it does not accumulate in the water. Damage results from the possible leakage of chemicals or oils, which are necessary for the operation of the vessel, not directly from the cargo content of the LNG tanker. From this perspective, the LNG tanker accident is comparable to any ship transporting cargo safe. Contrary, the part load is vaporized, it is estimated from 0.1% to 0.25% of total amount daily, it can be effectively used as fuel for the vessel. Thanks to that may be used up to 100% of this gas. (Chrz, 2009)

4 INLAND NAVIGATION - SOLUTION FOR COUNTRIES OF CENTRAL EUROPE

The most of the transport capacity of the current fleet of transoceanic ships carrying liquefied natural gas is made up of tankers with a capacity of 120,000 m^3 to 140,000 m^3. Construction of these ships is very complex and technologically demanding. Just only ten producers from all over the world have substantial experience with structures of this type. These include Finland (Kvaerner Masa), Germany (HDW), Italy (ItalcantieriGenoa, ItalcantieriSistri), France (Atlantique, La Ciotat, La Seine, La Trait), Japan (IHI Chita, ImabariHigaki, Imamura, Sakaide Kawasaki, Mitsubishi Nagasaki, NKK Tsu) North Korea (DaewooHanjin, Hyundai, Samsung), Netherlands (Bijlsma), Norway (MossMoss, MossStavanger), Spain (Astana, IZAR PuertoReal, IZAR Sestao), USA (GD Quincy).

Use of inland waterways for transportation of LNG is particularly relevant for landlocked countries of Central and Eastern Europe. Network of inland waterways of the European Union consists of approximately 37 000 km navigable rivers and canals. Interlinking Danube, Main and the Rhine by trans-European waterway was obtained connection of the Black and North Sea with a direct connection to a branched network of waterways of western France, Luxembourg, Switzerland, Germany and the Netherlands. This waterway has become one of the infra-structural priorities of European transport projects, taken within the European transport policy. The decisive goal of this priority is full of this important navigable waterway so that vessels can be transferred once as a group of goods from the North Sea to the Black Sea on the minimum weight of 3000 tons. Overall, the EU has earmarked for this task, the amount of 1 889 million € and from it 180 million € for the route Vienna – Bratislava. A significant amount is expected to use on the Lower Danube for removing ford sections with regard to the transport of heavy bulk items and also items containing dangerous cargo. An equally important activity for the Central European region in this direction is the effort to link the Danube with the North and Baltic Sea by canals and rivers Elbe and Oder. Czech, Slovak and Austrian investors, promote the implementation of project canal Danube - Oder - Elbe in the trans-EU and the European Agreement on main inland waterways of international importance. The aim of this project is connect the missing link in the waterway network and its implementation would allow countries of the region to maximize the gains from trade, including the extension of facilities for transportation of such commodities, such as LNG.

Vessels for LNG transportation by inland waterways have a capacity of 2000 - 4000 m^3, equivalent to 1.2 to 2.4 million m^3 of natural gas. Restrictions on the transport of liquefied natural gas associated with a sufficient bridges clearance on the waterway. Given the low density of LNG (0.45 t / m^3) issue draft of the vessel is negligible.

Maybe there is room for recovery in the recently neglected mixed river - sea technology, whose philosophy is based on the elimination of boundaries between sea and river, which means elimination of transhipment from marine vessels onto river vessels and back. The removing just one transhipment brings considerable economic and time savings. In this case there is no need to build (on the route) any pumping equipment and the ship can navigate from dispatch (liquefying terminal) to a port of destination. By conducted research can be concluded that the use of technology "river - sea" in comparison with separate technologies, "inland navigation" and "maritime navigation" is possible to reduce transport costs about 10% to 15%. Positive effects of this technology appear in connection with the organization of transport, particularly when they are introduced by providing logistical technological scheme "house to house".(Klepoch & Žarnay, 1998)

5 ROLE OF RIVER-SEA NAVIGATION IN THE EUROPEAN INLAND NAVIGATION SYSTEM

At the various international meetings relating to the further development of cooperation among the member countries of the Economic Commission for Europe (ECE) in the context of the AGN Agreement, attention is always given to the important role of river-sea navigation in developing the Pan-European inland navigation market. A number of studies suggest that the establishment of efficient coastal routes would have the following benefits: Transfer of foreign-trade freight traffic to river shipping; Completing the circle, currently broken in places, of category E waterways, linking the deep waterways of the European part of Russia to the network of European waterways of international significance and establishing a pan-European ring of trunk waterways around the whole of Europe; More effective use of the Rhine-Main-Danube trans-European trunk waterway and the pan-European transport corridors; Rendering transport operations more environmentally friendly and economically advantageous, since freight will be conveyed by inland waterways directly into the hinterland; Use of new transport and fleet management technologies and closer cooperation among the member countries of ECE in these matters; Promoting river-sea navigation on the waterways of France, Portugal, Spain and Italy.

The sea section of Don-Dnieper-Danube route is already widely used by Ukrainian and Russian combined river-sea navigation vessels, thanks to the favourable navigation and hydro meteorological conditions along the route during most of the year. Both in Ukraine and in Russia, river-sea vessels have basically been constructed in accordance with the class rules set down in the register of inland navigation vessels in the Russian Federation (the Russian River Register), although there are also a number of models of river-sea vessels which have been built to classes of the Russian maritime register and those of other classification societies.

Thought the closed circuit pan-European waterway system lays also at western part of Europe, the river-sea navigation does not have any tradition there, except in Netherlands. The most of classification organisations of Europe, Germanisher Lloyd, Norske Veritas or Buro Veritas does not have any vessel class designed for mixed river-sea navigation. They have very sophisticated system of classification, but only for river, or maritime vessels.

In the report of the standardization of ships and inland waterways for river-sea navigation, the Permanent International Association of Navigational Congresses (PIANC) recommended the following classes of vessels:

Tab. 1 Recommendation of basic dimensions of new conception river-sea vessels

River-sea class	Maximum permissible dimensions of vessels			Air clearance (m)
	Length (m)	Beam (m)	Draught (m)	
1	90	13	3.5 or 4.5	7 or 9.1
2	135	16	3.5 or 4.5	> 9.1
3	135	22.8	4.5	> 9.1

In fact, the Russian and Ukrainian vessel types listed above correspond fairly closely to those suggested by PIANC, although a draught of 4.5 metres

is unacceptable for the inland waterways along the route in question. Most of the river-sea vessels operated in the Russian Federation and Ukraine do not fully comply with all the height and draught limitations on certain waterways along the route of the future waterway ring around Europe. Accordingly, there is a need to develop new types of river-sea vessels with dimensions that meet the requirements for navigation both along the combined deep-water network of the European part of Russia and the Dnieper, and along the Rhine-Main-Danube route. (Klepoch & Žarnay, 2001)

6 CONCLUSIONS

Energy security is a key condition for the smooth functioning of states and is essential for the competitiveness of the economies of European countries. One of the primary energy sources is becoming a gas. Ensure its stable supply is one of the most contentious issues currently.

Europe has an extensive network of inland waterways that offer relatively inexpensive, efficient, clean and reliable mode of transport. Making more extensive use of LNG systems would enable European countries to take full advantage of the rapidly growing global market of natural gas, to make substantial long-term saving on their energy bill and to optimize storage and back-up capacities to compensate for shortages at peak times or to minimize energy supply shortfalls. Countries with well developed river and canal network, could envisage the development of LNG transportation to end users via inland waterways and thus creating a virtual network of pipelines, which avoids congestion and allows the LNG supply to urban areas, where geographical, demographic or environmental specificities are not suitable for the traditional laying of pipelines.

REFERENCES

Chrz, V. 2009. Informace k problematice bezpečnosti přepravy zkapalněného zemního plynu. In. *Vodní cestya plavba* 2: 28. Praha
Klepoch, J., Žarnay, P. 2001. Plavidlá pre dopravné technoló gie "rieka - more" In. *Komunikácie - vedecké listy Žilinskej univerzity = Communications – scientific letters of the Uni versity of Žilina*. č. 1 (2001), p. 103- 112.
Klepoch, J., Žarnay, P. 1998. The advance trends of "river – sea" transport Technologies enforcement in long distance traverse In: *Communications on the edge of the millenniums : 10th international scientific conference. 5th section, Quality and efficiency of transport, postal and telecommunications services. - Žilina : University of Žilina, 1998. - ISBN 80-7100-520-7. - p. 203-206.*
Melčák, M. 2010. Zlepšení energetické bezpečnosti Evropy vyšším využitím zkapalněného zemního plynu. In. *Parliamentary Assembly AS/EC(2010)09*

28. Novel Design of Inland Shipping Management Information System Based on WSN and Internet-of-things

Huafeng Wu, Xinqiang Chen, Qinyou Hu & Chaojian Shi
Merchant Marine College, Shanghai Maritime University, Shanghai, China

Jianying Mo
Science and Technology Division, Shanghai Maritime University, Shanghai, China

ABSTRACT: Currently there are more and more ships sailing in inland waterways so that the traditional inland shipping management information system (ISMIS) becomes relatively backward. Based on the rapidly developed new information technologies, such as wireless sensor network, Internet-of-things, cloud-computing and so on, we propose a novel design of ISMIS, which is featured by low cost, environment-friendly, cross platform, high scalability and integrity and thus can efficiently improve the inland shipping management and inland water environment.

1 INTRODUCTION

Wireless sensor networks (WSN) have attracted considerable amount of attention in recent years. There is a sort of sensors in WSN, including seismic sensors, temperature sensors, and humidity sensors etc. It is a new way of acquiring information platform that can supervise and collect various monitor objects' state. Its sensors can be positioned to many geographical areas that human being can hardly arrive even can't approach, for instance, volcano, the arctic pole and hostile battle fields and so on. Small size and minimum requirements from existing infrastructure make WSN one of revolutionary technologies in 21st century that will have a significant impact on our future life.

On the other hand, internet-of-things (IOT) is designed as a world-wide network in which everything can be identified by a unique address, every computer, each desk so much so that even a stone can acquire an exclusive address if we need. Every object can join the network dynamically and each terminal can collaborate and cooperate efficiently to achieve different tasks. IOT can realize real-time data obtaining, information exchange, remote control making use of traditional internet. IOT finds a range of applications in daily life including intelligent transportation system, smart home furniture, intelligent fireproof and so on.

Nowadays vessels sailing in inland waterways are increasing dramatically while the existing inland navigation management information system cannot provide more efficient service for inland rivers' management. Shortages of the management system have manifested themselves in following spheres.

Firstly, fewer applications of advanced techniques are employed in inland shipping administration. According to the survey statistics, ship-borne supervising devices are mostly deployed in ocean transportation vessels, namely AIS, GPS, RADAR, and other modern technologies are scarcely mounted in inland ships. Secondly, time-delay in current monitoring system is larger, in other words information or data broadcasted by the system are not latest. What's more it is difficult for such system to realize supervise and exchange information across the areas. As a result more efficient real-time shipping management system and more convenient data information sharing system are required. Development of WSNs and IOT make it possible to implement the very watercraft management system. The newly system can achieve low latency, inter-regional organizing network, easy realization and other advantages with the help of WSN and IOT for the aforementioned superiorities are the unique properties of these burgeoning IT technologies.

The remainder of the paper is organized as follows: In section 2 we present related work. In section 3, we propose whole design of inland shipping management system, including introduction of system functionality, design of network structure. Meanwhile system's data processing based on cloud computing is discussed in section 4 and middleware system of inland shipping management system is expounded in section 5. Then the paper is concluded in section 6.

2 RELATED WORK

I.F. Akyildiz et al. [1] provided a review of factors influencing the design of sensor networks and the communication architecture for sensor networks was outlined. In addition the algorithms, protocols for each layer and open research issues for the realization of sensor networks were also explored. He gives us a better understanding of applications of sensor network and the current research issues in the field, like node localization, designing energy efficient radio circuits, sensor network topology, fulfillment of its adaptivity to environment, protocols for sensor network's different layers and so on. It will be better if the authors talk about difference between traditional sensor networks, wireless sensor network as former can provide some assistance for the later.

The machine learning techniques applied in WSN from both networking and application perspectives were surveyed by Ma Di et al. [2]. Machine learning techniques have been applied in solving problems such as energy-aware communication, optimal sensor deployment and localization, resource allocation and ask scheduling in WSNs. In application domain, machine earning methods are mainly used in information processing such as data conditioning, machine inference and etc. The paper proposes a novel approach that first extends rigorously published mathematical constructs that merely approximate the long-term or stationary behavior of information flows in WSNs to simplify computation and simulation. Moreover, the approach, based on multivariate point processes, is shown to represent interaction of the parameters of the protocol layers by William S. Hortos [3]. Both Ma Di [2] and William S. Hortos [3] are talking about machine learning techniques while Ma focuses on the application aspects and William mainly on theoretic field, and their conclusions may be more stringency if they present some simulations.

When deploying several applications over the same WSN, One of the remaining problems resides in the data aggregation solutions, which are proposed generally for one application and may drain the WSN power in a multi-application context. Therefore, Ahmad Sardouk et al. [4] propose a data aggregation scheme based on a multi-agent system to aggregate the WSN information in an energy-efficient manner even if we are deploying several applications over this network. This proposal has proved its performance in the context of one and several applications through successive simulations in different network scales. Ahmad Sardouk gives us a concrete data priority processing scheme, and it would be better if it provides some mathematical model for the data aggregation scheme and discusses the scheme's node localization.

As ZigBee becomes a standard for WSN (Wireless Sensors Networks), and ZigBee and 802.15.4

had been proving they can achieve the mission splendid, Xavier Carcelle et al. [5] will emphasize on the past, present and future features for ZigBee, taking a look on the feedback from previous implementations to finally design the next generations of WSN based on ZigBee. Xavier Carcelle introduces the past, present and future features for ZigBee in detail, taking a look on the feedback from previous implementations to finally design the next generations of WSN applications based on ZigBee. If he connects WSN s' applications with some new developing techniques, such as internet-of-things, cloud computing and so forth, it will be perfect.

The research presented by Fuxing Yang et al. [6] has been focused on wireless gateway based on ZigBee and TD network. According to problems caused by limited bandwidth of traditional wireless sensory gateway, such as inferior network performance and low efficiency of communication, this paper comes up with a solution---a wireless sensory gateway based on ZigBee/TD, closely combining ZigBee net with TD-SCDMA net. In the plan of design, the network nodes transport data to gateway by ZigBee short-range communication technique after collecting them and the gateway sends them to control center by TD network from long distance, realizing the highly efficient long-range transport of data. The design satisfies the efficiency and transparency needed for inter-networks data exchange and it can be scalable easily. Fuxing Yang gives an excellent wireless gateway scheme and it may be more persuasive supposing that Fuxing Yang introduces some related work.

Miaomiao Wang et al.[7] provide a comprehensive review of the existing works on WSN middleware, seeking for a better understanding of the current issues and future directions in this field. They also propose a reference framework to analyze the functionalities of WSN middleware in terms of the system abstractions and the services provided and review the approaches and techniques for implementing the services. Based on the analysis and using a feature tree, the paper provides taxonomy of the features of WSN middleware and their relationships, and uses the taxonomy to classify and evaluate existing works. Open problems in this important area of research are also discussed in [7]. Miaomiao Wang gives us an overall detailed existing research on WSN middleware. The paper can be a more cogency survey as if he tells some transport protocols in WSN middleware.

Study on Internet-of-things is either in full swing. Jianhua Liu et al. [8] propose a formal IOT context model to perform self-adaptive dynamic service. They provide a general context aware service based on IOT communication and their context model is used for service match and service composition to reduce the consumption of devices resources and cost. Stephan Haller et al. [9] survey that how the Internet of Things is put in a wider context: how it relates to the Future Internet overall and where the business value lies so that it will become interesting for enterprises to invest in it. Stephan Haller also proposes the major application domains where the Internet of Things will play an important role and potential concrete business opportunities. Aitor Gomez-Goiri et al. [10] address the progress towards a semantic middleware which allows the communication between a wide range of embedded devices in a distributed, decoupled, and very expressive manner. This solution has been tested in a stereotypical deployment scenario showing the promising potential of this approach for local environments. Welbourne E. et al. [11] design a suite of Web-based, user-level tools and applications to empower users by facilitating their understanding, management, and control of personal Radio Frequency Identification (RFID) data and privacy settings. These applications are deployed in the RFID Ecosystem and a four-week user study is conducted to measure trends in adoption and utilization of the tools and applications as well as users' qualitative reactions.

Jianhua Liu [8] provides a simple enhanced dynamic service selection model and a formal internet-of-things (IOT) context model; both of the models are self-adaptive dynamic service model, and it would be better as if the paper provides some mathematical models. Stephan Haller et al. ([9]-[11]) tell us something more about IOT's concrete enterprises applications, combined with its specific applying techniques, such as RFID, middleware and so on. From personal point of view, the three papers need some simulations. Stephan Haller [9] need to supplement some content in chapter of the major technique issues, such as privacy, virtual and physical world fusion and so on, as these are important for the spread of IOT in future. Aitor Gomez-Goiri et al. [10] can introduce something concerned with IOT middleware and distributed computing so that the paper may be more consummate. Analogously, Welbourne E. [11] may replenish some information about RFID middleware for building the internet-of-things era.

3 INLAND SHIPPING MANAGEMNET INFORMATION SYSTEM

3.1 System functionality

Currently inland shipping management is in comparatively chaotic state. Vessel control and supervise departments are distributed in different regions such that management efficiency is lower for information barrier and information island. Evolution of IOT and WSN provides novel approach for internal navigation administration. Accordingly it is high time to establish a more efficient integrated management in-

formation system, which we refer to as Inland Shipping Management Information System (ISMIS) catering to the aforementioned claim.

ISMIS takes advantage of wireless sensor network technology, RFID technique, and IOT to provide real-time information for inland ships, vessel management department, and other correlative administrative departments. RFID tags embedded in ship can transform ships' dynamic and static information to base station, auxiliary sensors can be deployed in inland waterways, bridges, ships which haven't been equipped with RFID equipments or other identification facilities and other places if necessary. Wireless Sensor Networks can acquire and survey ships' state (including ship's name, vessel's number, tonnage of ship, vessel course, and navigational speed etc.) accurately such that upper computers and central control monitoring system can raise the management level tremendously.

WSN that used to detect environment and device status can avoid some unwanted accidents. For instance, sensors deployed in bridge will detect bridge's intensity of pier and bridge girder by the minute so that its hidden danger can be hustled out of the way as early as possible. Sensors in ISMIS can also aid navigation as they can broadcast fairway information ahead such as traffic condition, traffic density, navigational danger, water area state, and so forth.

What's more ISMIS provides a good basis for the standard of RFID technique, information system, information encoding, measurement, management mode, manipulation, and other technique touchstone.

3.2 Design of network framework

Network structure of ISMIS includes wireless sensor networks, cable network, terminal users and, management system server, and so forth. Cable network mainly presides over the communication among terminal customers, principal computers, and center controlling system. Data transmitted from base station or principal computers can be sent to center controlling system by cable network. Users' requests and administrators' management activities are executed through internet. Spot data and information delivered by RFID data acquisition unit can be gained by upper computers in real-time. The architecture of network frame is displayed in Fig. 1.

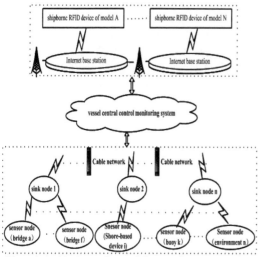

Figure 1. Architechture of network framework

Wireless sensor network in ISMIS primary implements function of detecting vessel information, environment state, bridge information, traffic status, and other factors which affect vessel operation. Sensor nodes can be divided into four categories: vessel nodes, bridge nodes, shore-based nodes, underwater nodes, and other nodes. With nodes distributed in a variety of monitoring areas, we can coverage most of supervising areas at the cost of lower costing such that all monitoring targets will transfer their status messages, at regular time, to base station or other data acquisition devices. WSN is ambient environment self-adaptively ad hoc network; it consists of physical layer, MAC layer, topology layer, network layer, transfer layer, and application layer. On one hand, RFID data collecting facilities sweep ship-borne RFID device to get watercraft's relative information at a fixed time, and data acquired in real-time are transmitted to local vessel management department and vessel center control monitoring system by RFID middleware system. On the other hand, vessel sensor nodes, which have been equipped with microchips that can realize information exchange automatically, underwater sensor nodes, shore-based nodes, and other sensor nodes shape ad hoc network. The ad hoc network is able to get and deliver supervising range's dynamic information and static state in real time which enhance remote management efficiency. Certainly WSN's information and data should be transferred to center management monitoring system with the help of WSN information middleware system. Partial pseudo-codes of WSN information middleware system are as follows:

```
Implementation {components vessel-node;
Class create-node-information
{Public:
Transfer (vessel-node information);
Private:
Char* vessel-node location;
Char* vessel-speed;
Char* vessel-direction;
int vessel-call-sign;
int vessel-IMO;
};
Refresh information database;
Return 0 ;}
```

3.3 Design of inland shipping management information system

Inland shipping management information system (ISMIS) employs wireless sensor network, RFID technique, IOT technology and other modern IT technique into inland vessel administration. ISMIS realizes precisely control over inland ships by the aid of WSN, IOT, and RFID, for the center control and management system of ISMIS can receive supervisory scope information concerning ships, navigation lock, fairway, traffic density, and other correlative data.

WSN in ISMIS plays the role of monitoring most targets, such as ships sailing in inland rivers, buoys, environment state, navigation danger, and so forth. WSN combines with RFID constitutes ISMIS bracket and network nerve. With the help of such modern IT technique, ISMIS can effectively schedule distributed vessel management subsystem and ship sailing in inland rivers so as to prevent traffic jam and accidents happening. Structure of ISMIS is showed in Fig. 2.

3.4 Achievement of dynamic node connecting to local cluster network

As vessels are shipping on inland rivers, nodes of WSN are changing constantly. Therefore locomotive ship nodes connect to local WSNs via wireless sensor networks gateway and concrete realization as follows:

Step 1: apply for joining native sub-WSN

A ship which functions as traveling node need to join native wireless sensor sub-network to formulate real time ad-hoc network. First thing for the mobile node is applying for registration in the sub-network. It broadcasts gateway request messages over the sub-network. The nearest sink node of the sub-network will build request-join node's path distance variable-Hops as soon as it receives registration messages delivered by ship node, and its initial value is 0. Then sink node sends permission messages of joining sub-network and application affirm

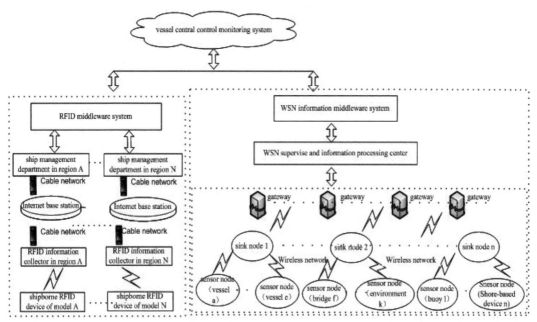

Figure 2. Architecture of inland shipping management information system

information to the application node by cluster's nodes. That very node will deliver response verification messages to sink node such that sink node can dynamic updates cluster nodes information.

Step 2: registration in the sub-network

Ship node transmits its relevant information, involving path hops, routing information and other information to sink node to login the cluster as a temporary member. Head-node forwards the receiving information to upper-computer system through wireless gateway. Upper- computer system generates provisional ID for new ship node and stores its information into ship node database. After that, extemporaneous ship node ID, registration identifying code and other essential data packets are sent to sink node and vessel sensor. Ship node can complete registering procedure so long as it receives extemporaneous ID and registration code. So far, sink node possess newcomer's information while it obtains upper-computer system's consent orders, such as registration node ID, registration area and so forth. Till then it is legal for the newcomer communicates with sink node and executes the task assigned by super-stratum node. The representation of the node joining local sub wireless sensor network is displayed in Fig. 3.

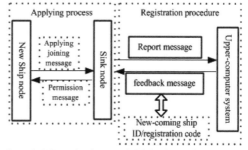

Figure 3. Schematic of new node joining into native sub-wsn

Step 3: task process

On the one hand, the new-coming member node needs to coordinate with sink node, shore-based nodes, environment supervising nodes and other nodes to realize information sharing that supervised and transmitted by upper nodes.

Ship-borne information collecting device, namely ship node transfers its data and acquired information to sink node which will transmit them to local vessel management department after data filtering and screening by network base station. Simultaneously, native environment nodes and bridge nodes and other relative nodes broadcast their monitoring data and messages to vessel controlling center which comprise of traffic density, water quality, embargo announcement, fair-way condition ahead and so on to shipping management centre to realize maximum efficiency of the whole wireless sensor network. In addition each sink node and nodes in every cluster communicate with each other through optimal path, while gateway link sensors are deployed in every navigation bridge, distributed to implement remote control and converting protocols between different networks and sink nodes.

4 DATA PROCESSING BASED ON CLOUD COMPUTING

Currently there is no uniform international definition of cloud computing. Domestic IT industry gives the definition of cloud computing as follows: cloud computing is the fusion of the traditional IT technologies and some newly developing techniques; the former involves grid computing, distributed computing, parallel computing, utility calculation, network storage, virtualization, load balance and the latter mainly comprises grid technology.

Inland shipping management information system as a large-scale inland crafts management system, it will handle, transmit, and store mass data as well as huge quantity of information and it really a challenge for we to implement sufficient facilities to accomplish such function. However cloud computing can meet our demand in acceptable overhead. For this reason we make use of could computing to conduct, store and broadcast our data as same as information.

The specific procedures as follows: firstly, data o messages delivered by RFID middleware system are decrypted, filtered, transformed in the light o stated format and requirement by distributed calculating center, which includes virtual computing centers and physical calculating centers, in cloud computing platform. Then vessel central contro monitoring system which is the command center o inland shipping management information systen will publish warning information for ships if any ob structions exist in front of the waterway. In addition cloud computing can also forward, convert format o correlative upper computers' orders for rock-botton devices and equipments. Besides, WSN middlewar systems dispatch their gathering information to ves sel center monitoring system through wireles transmission protocol and then all messages are con ducted by cloud computing which as with processing RFID devices' information.

5 MIDDLEWARE SYSTEMS DEVELOPMENT

In the inland shipping management information sys tem middleware systems are essential for the dat and information of ship-borne RFID devices and WSN nodes are incompatible in upper computer systems. Consequently we need to develop RFID middleware system and WSN middleware system respectively. Function modules of RFID middleware

system comprise formatting data, ensuring data communication security, data caching and filtering module, supplying application program interfaces etc [12]. That's to say, all data and messages delivered by ship-borne facilities will be processed before they are sent to vessel central control monitoring system. To the contrary, all messages would be transformed to compatible formats before they are transmitted to lower layer or bottom users broadcasted by ship central control supervisory system.

WSN middleware system is somewhat resembling RFID middleware system. The middleware system offers several terminal users' interfaces for outside applications, including ship node middleware interface, bridge node middleware interface, water-supervisory node middleware interface, and other comprehensive interfaces for different kind of sensor nodes. These interfaces will transmit and receive data or information as required in WSN processing and monitoring center. Internal function modules of WSN middleware will execute their regular work automatically as long as relative data or information access into database of middleware system. The architecture of WSN middleware system is shown is Fig. 4.

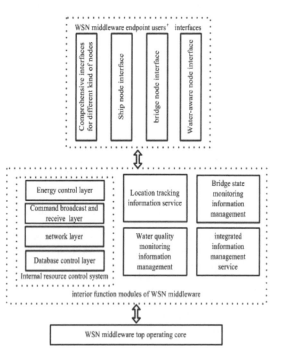

Figure 4. Architecture of WSN middleware system

6 CONCLUSION

This paper proposed a novel design of inland shipping management information system (ISMIS) for improving inland shipping management level. Since the design is based on wireless sensor network whose cost-effective is steadily higher the recent years, ISMIS can extend its supervising objects dynamically with the help of WSN, IOT, cloud computing, middleware technology and other IT techniques. Namely, inland shipping will be more efficient, real-time, and convenient than present inland shipping management system and inland ships will enjoy better information service in envisioned future with the establishment of ISMIS. In future we will pay more attention to the improvement and implementation of inland shipping intelligent management level using WSN and IOT techniques.

ACKNOWLEDGEMENT

This work was supported by Science & Technology Program of Shanghai Maritime University (20110028), Innovation Program of Shanghai Municipal Education Commission (09YZ247) and Shanghai Leading Academic Discipline Project (S30602).

REFERENCES

[1] I. F. Akyildiz, W. Su, Y. Sankara, subramaniam and E.Cayirci, "Wireless sensor networks: a survey," Computer Networks, vol. 38, pp. 393-422, Mar. 2002.

[2] Ma Di, Er Meng Joo, "A Survey of Machine Learning in Wireless Sensor Networks From networking and application perspectives," IEEE Information, Communications &Signal Processing, pp. 1-5, Dec. 2007.

[3] William S. Hortos, "Analytical Models of Cross-Layer Protocol Optimization in Real-Time Wireless Sensor Ad Hoc Networks," Lecture Notes of the Institute for Computer Sciences, Social Informatics, and Telecommunications Engineering, vol.28, pp.762-779, 2009.

[4] Ahmad Sardouk, Rana Rahim-Amoud, Leïla Merghem-Boulahia and Dominique Gaïti, "Data Aggregation Scheme for a Multi-Application WSN," Lecture Notes in Computer Science, vol.5842, pp.183-188, 2009.

[5] Xavier Carcelle, Bob Heile, Christian Chatellier and Patrick Pailler,"Next WSN applications using ZigBee," IFIP International Federation for Information Processing, vol.256, pp.239-254, 2008.

[6] Fuxing Yang, Chuansheng Yan, "Design of WSN Gateway Based on ZigBee and TD," Insitute of Electrical and Electronics Engineers, Inc, 2010 3th International Conference on Electronics and Information Engineering (ICEIE 2010), pp. v V2-76 - V2-80, 2010.

[7] Miaomiao Wang, Jiannong Cao, Jing Li, Sajal K. Dasi, "Middleware for Wireless Sensor Networks: A Survey,"Journal of Computer Science & Technology, vol. 23, pp 305-326, May. 2008.

[8] Jianhua Liu; Weiqin Tong, "Dynamic Service Model Based on Context Resources in the Internet of Things," Wireless Communications Networking and Mobile Computing (WiCOM), 2010 6th International Conference, pp. 1-4, 2010.

[9] Stephan Haller, Stamatis Karnouskos, Christoph Schroth, "The Internet of Things in an Enterprise Context," Computer science, vol. 5468/2009, pp 14-28, 2009.

[10] Aitor Gomez-Goiri, Diego Lopez-de-lpina, "A Triple Space-Based Semantic Distributed Middleware for Internet of Things," Computer Science, vol. 6385, pp 447-458, 2010

[11] Welbourne E., Battle L., Cole G., Gould K. etal, "Building the Internet of Things Using RFID:The RFID Ecosystem Experience," Internet Computing, IEEE, vol. 13, pp 48-55, May. 2009.

[12] Huafeng Wu, Xinqiang Chen, Qingyun Lian etc." Design of Cloud-computing-based Middleware for Internet-of-ships in Inland Shipping,"Insitute of Electrical and Electronics Engineers,Inc, 2010 3th International Conference on Power Electronics and Intelligent Transportation System, pp. 289-292, Nov.2010.

[13] LIU Lin-feng; LIU Ye, "Research on Wireless Sensor Network Architecture Model in Ocean Scene," Computer Science, vol. 37, pp. 74-77, Oct. 2010.

[14] MA Xiao Tie, LI Kai, "Application of wireless sensor network in human physical parameters collection," Microcomputer & Its Applications, vol. 29, pp. 39-42, 2010.

29. About Effectiveness of Complex Using of Satellite and Geoinformation Technologies on the Ship of Compound 'River-Sea' Type

A. Boykov
Department of Navigation, Moscow State Academy of Water Transport, Russia

V. Katenin
State Research Navigation and Hydrographic Institute, Russia

ABSTRACT: The problem of obtaining complete and accurate information for the skipper is crucial in making the right decisions to ensure safe navigation. Subject of study in this paper is the effectiveness of incurring way navigational watch with a wide variety of navigation equipment at the bridge. The authors demonstrate the benefits of integrated use of satellite & GIS technologies on ships to various options of using other electronic navigation systems.

SECTION I

Navigation security of navigation depend most on the opportuneness and basis of administrative solutions, which captain and his watch officer made during navigation watch, based on operative receipt complete and trustworthy information. However lack or abundance of such information in the extreme situations leads to the wrong actions with negative consequences.

The situation of the indefinity turned out in the case of information lack. That does not assist to acceptance of the correct decision.

In the case of the surplus of the information the situation is characterized that navigator can very difficulty detached main characteristic, to separate unessential, to reveal major correlations and make appropriate decision.

Thus contradictions arise between quantity and quality of the receipt information, ways of its remaking, reflection, keeping and using in the interests of navigation security. Producing contradictions provoke necessity of conducting special researches, which can make optimal decision through number and measure.

In accord to our opinion, that problem could be solved with the aid of complex using of new technologies in navigation: satellite system of different appointment and means of reflection of the heterogeneous information on the basis of geoinformation systems (GIS) which mean combination of computer equipment, program provision, geographical data and arbitrary project by user for accumulation, keeping, modification, processing, analysis and visualization of all forms of information.

It is a well know fact that digital map is a basis of any geographical information system.

Geoinformation system gives an opportunity to create maps in different scales and projections with different colouration, to define spatial links between maps objects, that is to create any necessary geographical maps which satisfied consumer's requests. In this way, the main appointment of GIS is an efficient presentation of reliable and processing space-dispersed information to the user which is necessary to solve administrative problems. It makes it irreplaceable analytical tool in daily and especially in extreme conditionals.

Success of GIS makes conditional on maintenance of the following main requests.

GIS must be:
- complete, that is inclusive all parties of information, program and technological provision which meet in the process of the system exploitation;
- complex, to give an opportunity of joint analysis of the large group of parameters in its correlation;
- open, to provide easiness of modification and resetting to maintain its on the level of contemporaneity, which is necessary as for providing evolutionary, so for solution of different tasks;
- covered, to provide protection of information which is intend for different administration levels.

GIS must provide solution of the following tasks:
- creation and conducting of the base space-dispersed data;
- creation and editing of digital maps in the different projections and scales;
- reflection of the different data in the form of maps, graphs, diagrams;
- analysis of mapping data;

- fathoming of geometrical characteristic of natural objects, distance from geographical points to the areas with either conditions;
- change of scales of reflection, forms and aspect of the presentation of graphic and maps information;
- binding to the information from data bases to geographical objects on the digital maps;
- interpolation and construction of vectorial and scalar fields on the information from data bases;
- fulfillment of inquiries on the different samples from data bases;
- on the reflection and spatial analysis of the maps data (on parameters, periods of time, regions, etc.);
- documentation of information production;
- application of supplement for carrying out special kinds of processing, keeping information, etc.

One of the complicated problems of the GIS-technology are the efficient gathering and keeping of initial data, which survey is the most labor-intensive and expensive process. Nowadays traditional sea and river paper navigation maps become the most widespread as the basis for creation of the digital maps. However in the case of the maps lack or necessity to get operative information of the large spatial inclusion using the distance means is the most expedient. The satellite information of the GIS has the special importance here. In the GIS the results of the distance investigation of the Earth surface (ocean) from the space are regularly renovative source of the data, which is necessary to form the information layers of the electronic maps in the large scales spectrum (from 1:10000 to 1:10000000). Information from the distance means of survey gives an opportunity not only to estimate efficiently, but mostly to renovate and correct using geographical stratum with exact arrangement of objects on them to geographical system of coordinate.

Including aforesaid, complex information from following satellite systems could be useful for forming geographical stratum:
- navigation (GLONASS, GPS, Gallileo, GLONASS+ GPS+ Gallileo);
- survey of situation;
- oceanographic;
- meteorological;
- geodesic.

Using of satellite – relay assists realization of connection between ships and command of different levels.

Integration of complex satellite information and GIS reveals new opportunities for providing navigation security of the navigation. New technology has following advantages which make it leader at the cost of:
- complex of information;
- operation of its receipt and presentation in compact form;
- graphic presentation;
- opportunity of analysis of joint heterogeneous information and production of well-founded administrative decisions.

To prove advantages of GIS before existing programmes of providing navigation security of sailing, based on using only electronic map and satellite navigation system could be possible using in the article new method of comparative estimate of effectiveness.

We could detached the following standard conditions of sailing and describe them briefly. According to the world practice regions of sailing separate on:
- oceanic where sailing accomplished behind boundary of the continent shelf or on the distance of more than 50 area miles from land or another obstacle;
- coastal where sailing accomplished in the boundaries of 50 sea miles from land or in the boundaries of external border of coastal bank or another dangers or in the area where sailing limited;
- ways to the harbours and sailing there, where sailing accomplished in the waters situated between land and area of the coastal sailing. Those regions are determined separately for each water way in the practice of navigation;
- internal water ways where sailing accomplished in limited areas, resemble with sailing in harbours or on the ways to its.

Therefore it is necessary to distinguish only 3 areas: oceanic, coastal and reduced sailing.

SECTION II

We could compare new technology with other approaches for comparative estimate of its effectiveness:
- without using of electronic maps;
- using electronic cartographical navigation-information systems for reflection of the definite itinerary of movement and reflection of necessary information for provision security of sailing.

As a criterion for comparing systems, we take the probability of obtaining reliable information for management decision $P_{ДОСТ}$ - multiplicative indicator species [4]

$$P_{НБП} = \prod_{i=1}^{n} P_i \qquad (1)$$

Where:
P_i – private dimensions (i=1,5);
P1 – accuracy rate sailing ship at any time;
P2 – measure the speed of information (РОП)- the probability of obtaining the necessary information in a specified time or in real time;
P3 – measure the validity of the information (РОБ) - the probability of obtaining the information that is adequate to the situation in which the

vessel is located. Calculated using mathematical modeling;

P4– measure the influence of the external environment (РГМО) - the probability of hydrometeorological information in real time, which affects the safety of navigation of the ship;

P5 – measure the influence of human factors on safety of navigation (РЧФ) - the probability of making correct management decisions in the light of human qualities and professional experience.

The primary means of determining the place of the ship are now satellite navigation systems are the second generation: GPS (U.S.) and GLONASS (Russia) and their functional additions.

Private index P1 is calculated as the probability of hitting a ship in a circle given radius or a strip of given width, respectively [5]

$$P_{\kappa} = 1 - \exp(-r/M)^2, \qquad (2)$$

$$P_{\pi} = \Phi\left(\frac{Ш_{\pi}}{M\sqrt{2}}\right) \qquad (3)$$

where
r - given radius;

M - experimental standard deviation determine of the position;

$$\Phi = \frac{2}{\sqrt{2\pi}} \int_0^x e^{-\frac{t^2}{2}} dt \qquad \text{- Laplace function.}$$

For the calculation of other partial indicators of efficiency can be used depending on the analytical or probabilistic estimates, which are an expert way [6,7].

Comparative analysis of the effectiveness of the proposed method is carried out for three options:
– option 1 - for the existing equipment from the navigating bridge
SNA receivers, but without the electronic cards;
– option 2 - for navigating bridge with the receivers of the SNA and electronic charts;
– option 3 - for navigating bridge with the receivers of various satellite systems and GIS-technologies.
The results of calculations by formula (1) are given in Tables 1-3.

Table 1. Existing equipment navigation bridge with the receivers of the SNA, but without the electronic cards

Watch	Navigation zones		
	Oceanic	Coastwise	Cramped conditions and GDP
Captain	P_{onep}=0,75 $P_{чф}$=0,97 P_{T}= 1,0 $P_{гмо}$=0,7 $P_{об}$=0,75 $P_{дост}$=0,38	P_{onep}= 0,8 $P_{чф}$=0,95 P_{T}= 1,0 $P_{гмо}$=0,75 $P_{об}$= 0,75 $P_{нбп}$=0,43	P_{onep}= 0,9 $P_{чф}$= 0,9 P_{T}= 1,0 $P_{гмо}$=0,85 $P_{об}$= 0,75 $P_{нбп}$=0,52
Senior Assistant	P_{onep}= 0,75 $P_{чф}$=0,9 P_{T}= 1,0 $P_{гмо}$=0,7 $P_{об}$=0,7 $P_{дост}$=0,33	P_{onep}= 0,8 $P_{чф}$=0,85 P_{T}= 1,0 $P_{гмо}$=0,75 $P_{об}$=0,7 $P_{дост}$=0,36	P_{onep}= 0,9 $P_{чф}$=0,8 P_{T}= 1,0 $P_{гмо}$=0,85 $P_{об}$=0,7 $P_{дост}$=0,43
Second Assistant	P_{onep}= 0,75 $P_{чф}$=0,8 P_{T}= 1,0 $P_{гмо}$=0,7 $P_{об}$=0,65 $P_{дост}$=0,27	P_{onep}= 0,8 $P_{чф}$=0,75 P_{T}= 1,0 $P_{гмо}$=0,75 $P_{об}$=0,6 $P_{дост}$=0,27	P_{onep}= 0,9 $P_{чф}$=0,7 P_{T}= 1,0 $P_{гмо}$=0,85 $P_{об}$=0,6 $P_{дост}$=0,32
Third Assistant	P_{onep}= 0,75 $P_{чф}$=0,7 P_{T}= 1,0 $P_{гмо}$=0,7 $P_{об}$=0,6 $P_{дост}$=0,22	P_{onep}=0,8 $P_{чф}$=0,7 P_{T}= 1,0 $P_{гмо}$=0,75 $P_{об}$=0,55 $P_{дост}$=0,23	P_{onep}= 0,9 $P_{чф}$=0,65 P_{T}= 1,0 $P_{гмо}$=0,85 $P_{об}$=0,55 $P_{дост}$=0,27

Table 2. Existing equipment navigation bridge with the receivers of the SNA and ECNIS

Watch	Navigation zones		
	Oceanic	Coastwise	Cramped conditions and GDP
Captain	P_{onep}=0,75 $P_{чф}$=0,97 P_{T}= 1,0 $P_{гмо}$=0,7 $P_{об}$=0,8 $P_{дост}$=0,41	P_{onep}= 0,85 $P_{чф}$=0,95 P_{T}= 1,0 $P_{гмо}$=0,75 $P_{об}$= 0,82 $P_{нбп}$=0,50	P_{onep}= 0,9 $P_{чф}$= 0,9 P_{T}= 1,0 $P_{гмо}$=0,85 $P_{об}$= 0,85 $P_{нбп}$=0,58
Senior Assistant	P_{onep}= 0,75 $P_{чф}$=0,9 P_{T}= 1,0 $P_{гмо}$=0,7 $P_{об}$=0,75 $P_{дост}$=0,35	P_{onep}= 0,85 $P_{чф}$=0,85 P_{T}= 1,0 $P_{гмо}$=0,75 $P_{об}$=0,77 $P_{дост}$=0,42	P_{onep}= 0,9 $P_{чф}$=0,8 P_{T}= 1,0 $P_{гмо}$=0,85 $P_{об}$=0,82 $P_{дост}$=0,50
Second Assistant	P_{onep}= 0,75 $P_{чф}$=0,8 P_{T}= 1,0 $P_{гмо}$=0,7 $P_{об}$=0,7 $P_{дост}$=0,29	P_{onep}= 0,85 $P_{чф}$=0,75 P_{T}= 1,0 $P_{гмо}$=0,75 $P_{об}$=0,72 $P_{дост}$=0,34	P_{onep}= 0,9 $P_{чф}$=0,7 P_{T}= 1,0 $P_{гмо}$=0,85 $P_{об}$=0,75 $P_{дост}$=0,40
Third Assistant	P_{onep}= 0,75 $P_{чф}$=0,7 P_{T}= 1,0 $P_{гмо}$=0,7 $P_{об}$=0,65 $P_{дост}$=0,24	P_{onep}=0,85 $P_{чф}$=0,7 P_{T}= 1,0 $P_{гмо}$=0,75 $P_{об}$=0,67 $P_{дост}$=0,30	P_{onep}= 0,9 $P_{чф}$=0,65 P_{T}= 1,0 $P_{гмо}$=0,85 $P_{об}$=0,70 $P_{дост}$=0,35

Table 3. Suspension bridge with the receivers of different Satellite Systems and GIS technologies

Watch	Navigation zones		
	Oceanic	Coastwise	Cramped conditions and GDP
Captain	$P_{опер}$=0,85 $P_{цф}$=0,97 $P_{т}$= 1,0 $P_{гмо}$=0,9 $P_{об}$=0,9 **$P_{дост}$=0,67**	$P_{опер}$= 0,9 $P_{цф}$=0,95 $P_{т}$= 1,0 $P_{гмо}$=0,95 $P_{об}$=0,92 **$P_{дост}$=0,75**	$P_{опер}$= 0,95 $P_{цф}$=0,9 $P_{т}$= 1,0 $P_{гмо}$=0,97 $P_{об}$=0,95 **$P_{дост}$=0,79**
Senior Assistant	$P_{опер}$= 0,85 $P_{цф}$=0,9 $P_{т}$= 1,0 $P_{гмо}$=0,9 $P_{об}$=0,9 **$P_{дост}$=0,62**	$P_{опер}$= 0,9 $P_{цф}$=0,85 $P_{т}$= 1,0 $P_{гмо}$=0,95 $P_{об}$=0,92 **$P_{дост}$=0,67**	$P_{опер}$= 0,95 $P_{цф}$=0,8 $P_{т}$= 1,0 $P_{гмо}$=0,97 $P_{об}$=0,95 **$P_{дост}$=0,70**
Second Assistant	$P_{опер}$= 0,85 $P_{цф}$=0,8 $P_{т}$= 1,0 $P_{гмо}$=0,9 $P_{об}$=0,9 **$P_{дост}$=0,55**	$P_{опер}$= 0,9 $P_{цф}$=0,75 $P_{т}$= 1,0 $P_{гмо}$=0,95 $P_{об}$=0,92 **$P_{дост}$=0,59**	$P_{опер}$= 0,95 $P_{цф}$=0,7 $P_{т}$= 1,0 $P_{гмо}$=0,97 $P_{об}$=0,95 **$P_{дост}$=0,61**
Third Assistant	$P_{опер}$= 0,85 $P_{цф}$=0,7 $P_{т}$= 1,0 $P_{гмо}$=0,9 $P_{об}$=0,9 **$P_{дост}$=0,48**	$P_{опер}$=0,9 $P_{цф}$=0,7 $P_{т}$= 1,0 $P_{гмо}$=0,95 $P_{об}$=0,92 **$P_{дост}$=0,55**	$P_{опер}$= 0,95 $P_{цф}$=0,65 $P_{т}$= 1,0 $P_{гмо}$=0,97 $P_{об}$=0,95 **$P_{дост}$=0,57**

Based on the outcome $P_{ДОСТ}$ taken from the tables, the graphs for the studied variants of the way and watch.

CONCLUSION

1 Common to all three cases is that the likelihood of obtaining reliable information increases for all members of the bridge watch on the transition from the oceanic area to the navigation of a ship sailing in cramped conditions and the GDP.
2 Using only SNA + ECNIS slightly increases the likelihood of obtaining reliable information $P_{ДОСТ}$. Sharp rise in $P_{ДОСТ}$ provides comprehensive use of heterogeneous satellite data and GIS technologies.
3 Under the first option (in the absence of electronic cartography and mapping of the external environment) there is the advantage of capital over other members of the bridge watch, especially in the area of constrained navigation. This advantages are:
  – from 1.15 times to 1.21 times over the senior assistant, respectively, in the ocean swimming area and swim in cramped conditions;
  – from 1.41 to 1.62 times over the second mate;
  – from 1.73 to 1.92 times over the third assistant
4 Under the second option (subject to availability of electronic means of cartography and mapping of the external environment) the advantage of capital over the rest of the way down to watch some of the same conditions and sailing is as follows:
  – from 1.17 to 1.16 times for the senior assistant;
  – from 1.41 to 1.45 times for the second assistant;
  – from 1.71 to 1.66 times for the third assistant.
  In this case, the greatest effect the introduction of electronic cartography and mapping of the external environment is achieved when navigating in the coastal zone.
5 Under the third option (subject to availability of information from different satellite systems and applications of GIS) is a significant reduction in the superiority of capital over the rest of the way the watch. In this case, the advantage of 1.1 times remains almost constant for all zones of navigation on the senior assistant, rising marginally from 1.22 times to 1.3 times for the considered zones of navigation on the second assistant, and slightly decreases from 1.4 times to 1.38 times over a third assistant to the same conditions of navigation.
6 When comparing the values $P_{ДОСТ}$ between variants may be noted
7 The biggest advantage of the integrated use of heterogeneous satellite data and GIS-technologies received the third mate in the area of coastal vessels.
  The smallest advantage of this complex technology to access capital when navigating in restricted conditions.
8 Application of satellite and GIS technologies, to some extent negates the professional experience and expertise of the bridge watch ($P_{об}$). Thereby reducing the risk of human factors on the wrong decision and created considerable promise for automating the navigation process.

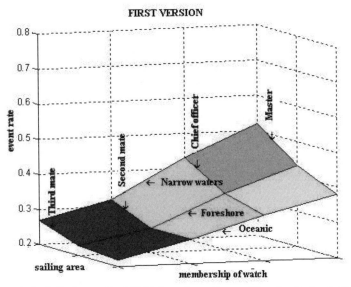

Figure 1. Comparative analysis of effectiveness ($P_{дост}$) Watch for various sailing conditions for a hardware version of the navigation bridge

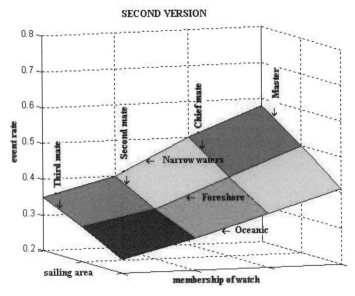

Figure 2. Comparative analysis of effectiveness ($P_{дост}$) watches in different sailing conditions for the 2 variants of the navigation bridge equipment

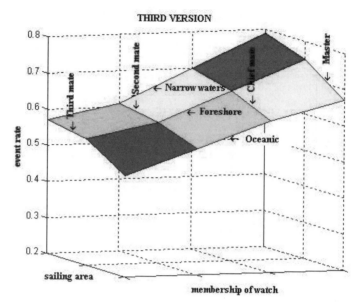

Figure 3. Comparative analysis of effectiveness ($P_{дост}$) watches in different sailing conditions for the 3 types of equipment navigating bridge

REFERENCES

[1] Tikunov, V.S. 1997. Modeling in cartography. Moscow: MGU.
[2] Katenin, V.A. & Katenin, A.V. 2002. New features integrated use of satellite and GIS technologies for navigation and hydrographic support. Information and Control Systems 1: 18-23.
[3] IALA: Guidance on navigational equipment (NAVGUIDE). 2001. St.-Peterburg: GUNiO.
[4] Kondrashihin, V.T. 1989. Determination of place the vessel. Moscow: Transport.
[5] Ventcel, E.S. & Ovcharov, L.A. 1988. Probability theory and its engineering applications. Moscow: Nauka.
[6] Martino, J. 1977. Technological forecasting. Moscow: Progress.
[7] Beshelev, S.D. & Gurvic, F.G. 1980. Mathematical and statistical methods of expert estimates. Moscow: Statistica

30. SC-Method of Adaptation Marine Navigational Simulators for Training River Shipmasters

O. Demchenkov
Moscow State Academy of Water Transport, Russia

ABSTRACT: The article provides an actual overview of navigational simulators for training shipmasters to control vessels on inland waterways in Russian Federation. There're considered methodology of adaptation "marine" navigational simulators for training "river" shipmasters, based on SC-method ("the Safety Cube" method).

INTRODUCTION

Among the most important tasks of the professional training for the interior water transport specialists are the safety of navigation and the decrease in the number of accidents on the inland waterways. This is achieved largely by the up keeping of knowledge, skills and practice of the navigators at an appropriate level. The navigational simulators have been the main tools for training and monitoring the vessel navigator's level of skill. In spite of the measures to enhance the professional level of the inland water transport specialists, the number of accidents due the vessel navigator's mistakes is not diminishing. According to data provided by the RF State Organization for Naval and River Navigation Supervision from 2005 to 2010 (see ref.1), the accident number remains at the same level and even has the trend to increase. This fact indirectly shows a low effectiveness of the existing methods of the navigator's professional training.

BASIC CONCEPT OF THE SC-METHOD

The analysis of the recent publications on the problems of increasing the navigators qualification level (see ref.2,3) permits to draw the conclusion about the need to improve in the fast place the methods of using the navigational simulators to train the inland water transport navigators. At present time there is no in Russia common concept of using the navigational simulators for the river vessel masters training. The training with simulators has been envisaged only for the "river-marine" vessel navigators. As for the other categories of inland water transport navigators, the right to choose the mode of their training has been delegated to the leaders of particular water

basins and steamship companies. Therefore, the selection of navigational simulators to be used in training centers depends in a large degree on the competence of decision taking leaders and on the financial prosperity of organizations. Thus, there is a real need in developing clear criteria of evaluating the effectiveness of simulators, produced by different companies to be employed in training specialists for specific water basins of the Russian Federation.

In order to substantiate the optimum criteria of employing simulators for the purpose of training the specialists of inland water transport, the author have analyzed in the period of 2009 to 2010 the database of the simulators used in 33 training centers of the RF inland water transport. In the course of research the author has taken into consideration the requirements contained in guidance documents (see ref.4-7) as well as the published materials (see ref.8,9).

In general, the number of parameter groups that can be used to compare the simulators, may be rather large. At the same time, any systematic apparatus should be developed in the interests of its practical use and must be easily understood by the specialists having a different level of mathematical knowledge. Therefore, in the process of developing the methods for the efficiency evaluation for the navigational simulators used in training of vessel navigators of the inland water transport, the basic data was grouped according to the following three categories:

1. transport infrastructure facilities (i.e. vessels, hydrotechnical constructions, distinctive parts of the inland waterways;
2. water basins;
3. navigational simulators used for the training of inland water transport navigator

Table 1. Typical transport infrastructure facilities of real basins from the RF inland waterways. XY-projection according to SC-method

Water basins	Vessels								Hydrotechnical constructions						Distinctive parts of the inland waterways										Bridges			
															Free rivers (rivers without obstacles)						Rivers with gateways							
	Self-propelled ships	Haulted ships	Kicked ships	Passenger carriers*	Dangerous goods carriers	Rapid ships	Barges	Rafts	Suction-tube dredgers	Single-ribbed locks	Double ribbed locks	Single-chambered locks	Multi-chambered locks	Moorings	Group rifts	Rifts between riversides	Rifts with placers	Rifts along ravines	Abrupt turns	Lateral channels	Lake segments of the water basin	River segments of the water basin	Lakes	Channels	Single-span bridges	Multi-span bridges	Areas with avanport	Seasonal prevalence**
Azovo-Donskoy	+	+	+	+	+	+	+	−	+	+	+	+	−	+	+	+	+	+	+	+	+	+	−	+	+	+	+	+
Amursky	+	+	+	−	+	+	+	−	+	−	−	−	−	+	+	+	+	+	+	+	+	+	−	−	−	+	−	+
Volzhsky	+	+	+	+	+	+	+	−	+	+	+	+	+	+	+	+	+	+	+	+	+	+	−	−	+	+	+	+
Vostochno-Sibirsky	+	+	+	−	+	+	+	+	+	+	+	−	−	+	−	+	−	+	+	+	−	+	+	−	+	+	−	−
Enisejsky	+	+	+	−	+	+	+	+	+	+	+	−	−	+	−	+	+	+	+	+	−	−	−	−	+	+	−	−
Kamsky	+	+	+	+	+	+	+	+	+	+	+	+	−	+	+	+	−	+	+	−	+	+	−	+	+	+	+	+
Severo-Vostochny	+	+	+	+	+	+	+	+	+	+	−	−	−	+	−	+	+	+	+	+	+	+	+	+	+	+	+	+
Severo-Zapadny	+	+	+	+	+	+	+	+	+	+	−	−	+	+	−	+	−	+	+	+	+	+	+	+	+	−	+	−
Severny	+	+	+	−	+	+	+	+	+	+	−	−	−	+	+	+	+	+	+	+	+	+	−	−	+	+	+	+
Obsky	+	+	+	−	+	+	+	+	+	+	−	+	+	+	−	+	+	+	+	+	−	+	−	+	+	+	+	−
Ob'-Irtyshsky	+	−	+	−	+	+	+	+	+	+	−	+	+	+	−	+	+	+	+	+	−	−	−	+	+	+	+	−
Centralny	+	+	+	+	+	+	+	+	+	+	+	+	+	+	+	+	+	+	+	+	+	+	−	+	+	+	+	−

* Considered 3-screws passenger ship with a separate management for machine (screw).

** Under the seasonal prevalence understood multifactorial concept that consists of: changes in the position and direction of the fairway due to due to rearrangement of sediments; changes radar images in connection with water discharge in the upper reach of the reservoir; изменение скорости и направления течения; change in the depth of the fairway, and other factors..

Table 2. Transport infrastructure facilities which are modeled in navigational simulators. XZ-projection according to SC-method

Navigational simulators	Vessels								Hydrotechnical constructions						Free rivers (rivers without obstacles)						Rivers with gateways				Bridges			
	Self-propelled ships	Hauled ships	Kicked ships	Passenger carriers *	Dangerous goods carriers	Rapid ships	Barges	Rafts	Suction-tube dredger	Single-ribbed locks	Double ribbed locks	Single-chambered locks	Multi-chambered locks	Moorings	Group rifts	Rifts between riversides	Rifts with placers	Rifts along ravines	Abrupt turns	Lateral channels	Lake segment of the water basin	River segment of the water basin	Lakes	Channels	Single-span bridges	Multi-span bridges	Areas with avanport	Seasonal prevalence **
«NTPro»	+	-	+	-	+	+	+	-	-	-	-	+ ***	-	+	-	+	-	+	+	+	-	+	+		+	+	+	-
«MARLOT»	+	-	-	-	+	-	-	-	-	-	-	-	-	+	-	-	-	-	+	+	-	+	+	+	+	+	-	-
«MASTER»	+	-	+	-	+	-	+	-	-	-	-	-	-	-	-	-	-	+	+	+	-	+	+	+	+	+	-	-
«RNM»	+	-	+	-	+	-	-	-	-	+	-	+	-	+	-	-	-	-	+	-	-	+	+	+	+	-	-	-
«Riv.Sim. 2.5»	+	-	+	-	-	-	-	-	-	-	-	-	-	-	-	-	-	+	+	+	-	+	+	-	+	+	-	-

---

* Considered 3-screws passenger ship with a separate management for machine (screw).

** Under the seasonal prevalence understood multifactorial concept that consists of: changes in the position and direction of the fairway due to due to rearrangement of sediments; changes radar images in connection with water discharge in the upper reach of the reservoir; изменение скорости и направления течения; change in the depth of the fairway, and other factors..

*** Gateway as an object of transport infrastructure not realized in full in "NTPro". Only entry or exit from the gateway are modeled without the implementation of the locking process.

Table 3. Navigational simulators which are used for training shipmasters to control vessels in real basins from the RF inland waterways. ZY-projection according to SC-method

Water basins	Navigational simulators				
	«NTPro»	«MARLOT»	«MASTER»	«RNM»	«Riv.Sim. 2.5»
Azovo-Donskoy	+	-	+	-	-
Amursky	+	-	-	-	+
Volzhsky	+	+	+	+	-
Vostochno-Sibirsky	+	-	-	-	+
Enisejsky	+	-	-	-	+
Kamsky	+	-	-	+	-
Severo-Vostochny	+	-	-	-	+
Severo-Zapadny	+	+	-	+	-
Severny	+	+	-	+	-
Obsky	+	-	+	+	-
Ob'-Irtyshsky	+	+	+	+	-
Centralny	-	+	+	-	-

The mathematic models realizing method called "the Safety Cube" (SC-method) (see ref.10) are the best suitable for the solution of multi-criteria optimization tasks of similar class. This approach consists in establishing interconnection between the constituent elements (components) of researched categories and comes to the construction of developments along the axes XY, XZ and ZY (see fig.1)

Tables 1, 2 and 3 contain the detailed characteristics of interrelation of the components of the researched categories. They are grouped according to the evolvent XY, XZ and ZY correspondingly.

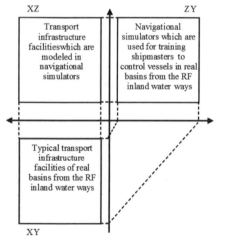

Figure1. The interrelation of categories used for evaluating the effectiveness of navigational simulators, using the SC-methodology apparatus

USING THE SC-METHOD TO ASSESSING EFFECTIVENESS OF MARINE NAVIGATIONAL SIMULATORS FOR TRAINING RIVER SHIPMASTERS

A detailed analysis of evolvents of "the Safety Cube" allows detecting individual features of navigational simulators which are used in training centers for training navigators. Knowledge the differences and shortcomings of navigational simulators from different manufacturers may help decision makers (managers of shipping companies, leaders of basins, etc.) to plan effective teaching and/or retraining of crew for the operation vessels in specific inland water basins.

In particular, using the above evolvents can be convincingly argued that in modern navigational simulators the transport infrastructure facilities such as vessels and hydrotechnical constructions (waterworks) are presented to a limited amount and does not fully reflect the real diversity of existing types of vessels and waterworks of the Russian Federation inland waterways. Also, a number of simulators, which was originally designed to prepare the skippers of marine vessels are absent or are not modeled in full amount typical parts of inland waterways (rifts, canals, etc.).

Except qualitative assessments, using evolvents of "the Safety Cube", there is possibility to obtain quantitative expert estimates. For example, in Table 4 shows the percentages of implementation the transport infrastructure facilities of RF inland waterways in navigational simulators with standard kit, obtained from the results of statistical processing the information from the evolvent XZ.

Table 4. Percentages of implementation the transport infrastructure facilities of RF inland waterways in navigational simulators with standard kit

Navigational simulators for RF training centers	Percentage of implementation the transport infrastructure facilities in navigational simulators			
	Vessels	Hydro-technical constructions	Distinctive parts of the inland waterways	In total
«NTPro»	55.5%	20%	64.3%	53.6%
«MARLOT»	22.2%	20%	50%	35.7%
«MASTER»	44.4%	0%	57.1%	42.8%
«RNM»	22.2%	60%	35.7%	35.7%
«Riv.Sim. 2.5»	22.2%	0%	35.7%	25%

For clarity, the data of Table 4 can be represented graphically by plotting values of the calculated parameters on the axes of the polar coordinate, as shown in Figure 2.

The data in Table 4 allow confirming with quantitatively mentioned above qualitative conclusion about the absence of a complete list of transport infrastructure facilities in modern navigational simulators which are used for training and retraining shipmasters and navigators for inland waterways

On average over all navigational simulators in Russian training centers, are able to simulate only 38.6% of objects from inland water transport infrastructure! In turn, officials of shipping companies which are responsible for the safety of navigation can make a reasoned conclusion about the low efficiency of use of modern marine navigational simulators for training river shipmasters, intended for testing skipper's skills in different shipping conditions on inland waterways.

CONCLUSIONS

Thus, the SC-method as a variation of methodical apparatus "the Safety Cube" really allows implementing a way of evaluating the effectiveness of the navigational simulators for training shipmasters to control vessels on inland waterways in Russian Federation. Thus, the introduction of SC-method will contribute costs optimization to simulators, will increase the quality of training navigators by rules for necessary adaptation and eventually will decrease the number of accidents and crashes on the inland waterways.

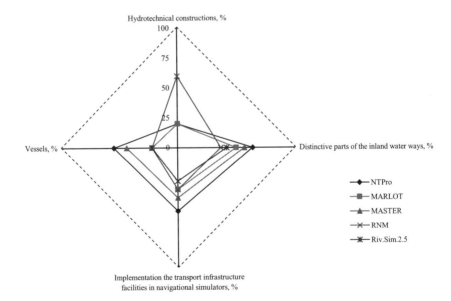

Figure 2. Percentage of implementation the basic groups of the transport infrastructure facilities in navigational simulators with standard complectation

At present time is ready a series of publications how to use proposed in article the SC-method for rationale selecting concrete model and modification of navigational simulators for training navigators for specific river basins of the inland waterways of Russian Federation. Now is developed some practical methodics based on the SC-method for forming requirements for databases of river navigational simulators, taking into account regional specificities, for the following organizations: Bashkir River Shipping, Shipping Company "ORION", Research Company "Systems&Technologies" and Kotlas Rivership College.

REFERENCES

[1] Official website of the Office of the State of Maritime and River Oversight (Gosmorrechnadzor)
URL: http://www.rostransnadzor.ru/sea/

[2] Aizinov, S.D. 2006. Analysis of the effectiveness of marine simulators. Morskoy flot (the Sea Fleet) 6: 18–23.

[3] Kostylev, I.I. 2006. Status and prospects of development of simulator training the specialists of maritime transport. Morskoy flot (the Sea Fleet) 6: 8 – 14.

[4] Instruction by content navigational equipment of inland waterways / General Directorate of waterways and waterworks Minrechflota RSFSR. 1988. Moscow: Transport.

[5] the Code of Inland Water Transport №374-FZ. 2009.

[6] Rules for navigation on inland waterways of the Russian Federation. 2009. Moscow: Translit

[7] Rules for crossing vessels and convoys through the locks of inland waterways of the Russian Federation. 2004. Moscow: Translit.

[8] Golubev, A.I. 1987. Radar methods of navigation on inland waterways. Moscow: Transport.

[9] Katenin, V.A.& Zernov, A.V.& Fadeev G.G. 2010 Navigation and hydrographic support on the inland waterways. Moscow: Morkniga.

[10] Japparov, E.R. 2009. Methodical apparatus preventive rationing technogenic transport risks on inland waterways of the Russian Federation. Moscow: Altair-MGAWT

Author index